Ergebnisse der Mathematik
und ihrer Grenzgebiete Band 84

Herausgegeben von P. R. Halmos P. J. Hilton
R. Remmert B. Szökefalvi-Nagy

Unter Mitwirkung von L. V. Ahlfors R. Baer
F. L. Bauer A. Dold J. L. Doob S. Eilenberg
K. W. Gruenberg M. Kneser G. H. Müller
M. M. Postnikov B. Segre E. Sperner

Geschäftsführender Herausgeber P. J. Hilton

J.H. Wells   L.R. Williams

Embeddings
and Extensions in Analysis

Springer-Verlag
Berlin Heidelberg New York 1975

James H. Wells
University of Kentucky, Lexington, U.S.A.

Lynn R. Williams
Louisiana State University, Baton Rouge, U.S.A.

AMS Subject Classifications (1970):
42 A 72, 42 A 88, 44 A 10, 52 A 40, 52 A 45, 52 A 55, 54 C 25, 54 E 40

ISBN-13:978-3-642-66039-9   e-ISBN-13:978-3-642-66037-5
DOI: 10.1007/978-3-642-66037-5

Library of Congress Cataloging in Publication Data.
Wells, James Howard, 1932-. Embeddings and extensions in analysis. (Ergebnisse der Mathematik und
ihrer Grenzgebiete; Bd. 84). Bibliography: p. Includes index. 1. Topological imbeddings. 2. Isometrics
(Mathematics). 3. Metric spaces. 4. L^p spaces. I. Williams, Lynn R., 1945- joint author. II. Title. III. Series.
QA611.21.W44. 514′.3. 74-31234

© by Springer-Verlag Berlin Heidelberg 1975
Softcover reprint of the hardcover 1st edition 1975

Preface

The object of this book is a presentation of the major results relating to two geometrically inspired problems in analysis. One is that of determining which metric spaces can be isometrically embedded in a Hilbert space or, more generally, in an L^p space; the other asks for conditions on a pair of metric spaces which will ensure that every contraction or every Lipschitz-Hölder map from a subset of X into Y is extendable to a map of the same type from X into Y.

The initial work on isometric embedding was begun by K. Menger [1928] with his metric investigations of Euclidean geometries and continued, in its analytical formulation, by I. J. Schoenberg [1935] in a series of papers of classical elegance. The problem of extending Lipschitz-Hölder and contraction maps was first treated by E. J. McShane and M. D. Kirszbraun [1934]. Following a period of relative inactivity, attention was again drawn to these two problems by G. Minty's work on non-linear monotone operators in Hilbert space [1962]; by S. Schönbeck's fundamental work in characterizing those pairs (X,Y) of Banach spaces for which extension of contractions is always possible [1966]; and by the generalization of many of Schoenberg's embedding theorems to the setting of L^p spaces by Bretagnolle, Dachuna Castelle and Krivine [1966].

The dominant theme in the subject of isometric embedding is that of positive definite quadratic forms and radial positive definite functions on metric spaces; in the problem of extending contractions the pervasive idea is an intersection property for spheres due to Kirszbraun which, in its simplest form, asserts that if a family of circular discs in the plane having a point in common are displaced so that their centers are no farther apart than before, then they again have a point in common. With certain pairs of metric spaces the Kirszbraun intersection property can be interpreted in the context of a procedure for solving systems of quadratic inequalities. These K-functions, as we shall call them, are at the heart of the extension problem for the class of Lipschitz-Hölder maps of order α between L^p spaces and they are structurally equivalent to the quadratic forms associated with the embedding of metric spaces in Hilbert space. Thus, certain aspects of the two subjects focus on a common class of quadratic forms, and it is this interaction that first kindled our interest and motivated this unified treatment.

Almost all the results we include have appeared in the literature. But as in any work of this type we have chosen to revise some proofs and include others heretofore unpublished. We make no claim of encyclopedic coverage, for we have exercised the usual perogative of concentrating on those topics which seem to us most interesting and significant. Nonetheless we have attempted to provide the reader with a bibliography which will serve as an adequate guide to the history and the current status of these subjects.

This book is an outgrowth of a seminar given by the first-named author at Texas Tech University and, at a crucial stage, its progress was stimulated through a sabbatical leave granted by the University of Kentucky. Thanks are due to Professor T. L. Hayden who introduced us to these subjects and to Professor A. L. Shields who pointed out the relationship between our work and certain packing problems. And thanks are also due to Wanda Jones, who typed the manuscript.

Table of Contents

Chapter I. Isometric Embedding

§1. Introduction

In this chapter we treat two special cases of the general geometric problem of determining those metric or linear spaces which can be isometrically embedded in a given Banach space. First, we consider the question of which metric spaces (X,ρ) can be isometrically embedded in a Hilbert space H, that is, under what metric conditions does there exist a map $\phi: X \to H$ such that

$$\|\phi(s) - \phi(t)\| = \rho(s,t) \tag{1.1}$$

for all points s and t in X? Secondly, we characterize those real normed linear spaces which are linearly isometric to a subspace of some L^p space, $1 \leq p \leq 2$.

Within the context of topological embedding the notion of isometric embedding is admittedly restrictive; however, as we shall see in the sequel, it is precisely this feature which permits the introduction of analytical methods. Also, as classical results show, the problem posed is radically altered if L^p space is replaced by a general Banach space or if the embedding is merely required to be topological. Specifically, a result of Banach and Mazur [4, p.187] states that any separable metric space can be isometrically embedded in $C[0,1]$, the space of real continuous functions on the interval $[0,1]$. More generally, every metric space can be isometrically embedded as a closed subset of some normed linear space [2]. And a special case of a well known theorem of Urysohn shows that any separable metric space can be topologically embedded in Hilbert space. Thus the separable Banach space ℓ^p ($1 \leq p < \infty$, $p \neq 2$) can be topologically embedded in Hilbert space—but as we point out in §2, isometric embedding is impossible.

§2. Isometric Embedding in Hilbert Space

Throughout H will denote a real Hilbert space in which inner product and norm are indicated by the symbols $(\cdot,\cdot)$ and $\|\cdot\| = (\cdot,\cdot)^{1/2}$. For each positive integer n, $\mathbb{R}^n$ denotes n-dimensional Euclidean space with the usual inner product and norm: $(x,y) = \Sigma_1^n x_i y_i$ and $\|x\| = (x,x)^{1/2}$. We shall view $\mathbb{R}^n$ as a subspace of $\mathbb{R}^m$ when $n < m$ and, when convenient, consider any one of these spaces as being isomorphically isometric and hence identical with a subspace of H of the appropriate dimension.

The words "embedding" and "isometric embedding" will be used synonymously and, as the context will indicate, ρ is either a metric, a semi-metric or a quasi-metric for the set X.

Most of the results in this section are taken from Schoenberg's papers [77] and [80].

Theorem 2.1. *The finite metric space* $\{x_0,x_1,x_2,...,x_n\}$ $(n > 2)$ *is embeddable in* $\mathbb{R}^n$ *if and only if the quadratic form*

$$\sum_{j,k=1}^{n} \tfrac{1}{2}[\rho(x_0,x_j)^2 + \rho(x_0,x_k)^2 - \rho(x_j,x_k)^2]\, \xi_j\xi_k \tag{2.1}$$

is positive semidefinite, that is, ≥ 0 *for all choices of real numbers* $\xi_1,\xi_2,...,\xi_n$.

Proof. If embedding is possible there exist $n+1$ points $y_0,y_1,...,y_n$ in $\mathbb{R}^n$ such that

$$\|y_j - y_k\| = \rho(x_j,x_k) \qquad (0 \leq j,\, k \leq n), \tag{2.2}$$

and it is clearly permissible to take $y_0 = 0$. If we replace $y_j - y_k$ by $(y_j - y_0) + (y_0 - y_k)$, square both sides and expand on the left using inner products, (2.2) takes the form

$$(y_j,y_k) = \tfrac{1}{2}[\rho(x_0,x_j)^2 + \rho(x_0,x_k)^2 - \rho(x_j,x_k)^2]. \tag{2.3}$$

That (2.1) is positive semidefinite now follows upon making the obvious substitution in the inequality

$$0 \leq \Big\|\sum_{1}^{n} \xi_j y_j\Big\|^2 = \sum_{j,k=1}^{n} (y_j,y_k)\, \xi_j\xi_k.$$

For the sufficiency, suppose that the quadratic form (2.1) is positive semidefinite. This is equivalent to the assertion that the real symmetric $n \times n$ matrix $A = (a_{jk})$, where

$$a_{jk} = \tfrac{1}{2}[\rho(x_0,x_j)^2 + \rho(x_0,x_k)^2 - \rho(x_j,x_k)^2] \quad (1 \leq j,\, k \leq n), \tag{2.4}$$

is positive semidefinite on $\mathbb{R}^n$. Thus, letting A also denote the corresponding linear transformation, we have

$$(A\xi,\xi) \geq 0 \qquad (\xi = (\xi_1,\xi_2,...,\xi_n) \in \mathbb{R}^n);$$

and therefore there exists a real nonsingular matrix D and a positive integer r, $1 \leq r \leq n$, such that if $\xi \in \mathbb{R}^n$ and $\eta = D\xi$, then

$$(A\xi,\xi) = \eta_1^2 + \eta_2^2 + \cdots + \eta_r^2 = \|P_r\eta\|^2. \tag{2.5}$$

The integer r is the rank of the matrix A and P_r is the orthogonal projection of $\mathbb{R}^n$ onto $\mathbb{R}^r$.

Denote the standard basis in $\mathbb{R}^n$ by $e_1,e_2,...,e_n$, set $y_j = De_j$, $1 \leq j \leq n$, and then successively substitute e_j and $e_j - e_k$ for $\xi = (\xi_1,\xi_2,...,\xi_n)$ in (2.5) to obtain the relations

$$\rho(x_0,x_j)^2 = (Ae_j,e_j) = \|P_r y_j\|^2 \quad (1 \leq j \leq n),$$

$$\rho(x_j,x_k)^2 = (A(e_j - e_k),\, (e_j - e_k)) = \|P_r(y_j - y_k)\|^2 \quad (1 \leq j,\, k \leq n).$$

Hence the map that sends x_0 to 0 and x_j to $P_r y_j$ defines an isometry from $\{x_0,x_1,...,x_n\}$ into R^r, and the proof is complete. $\square$

Inspection of this proof, with particular attention to (2.5), reveals the following corollary.

Corollary 2.2.   *A finite metric space $\{x_0,x_1,\dots,x_n\}$ can be isometrically embedded in Euclidean space $\mathbb{R}^r$, $1 \leq r \leq n$, of dimension r and no smaller dimension, if and only if the quadratic form (2.1) is positive semidefinite and of rank r.*

If an infinite metric space (X,ρ) is to be isometrically embeddable in Hilbert space H, then certainly each of its finite subsets must satisfy the quadratic form condition (2.1). And it is not difficult to see that (2.1) is also sufficient for embedding in case X is separable. For then one can choose a countable dense subset $S = \{x_0,x_1,x_2,\dots\}$ of X and construct isometries ϕ_n from $S_n = \{x_0,x_1,\dots,x_n\}$ into H in such a way that ϕ_{n+1} is an extension of ϕ_n, $n = 1,2,\dots$. This allows one to define an isometry on S and then extend it to all of X by continuity. For a general metric space the proof is somewhat less instructive.

Lemma 2.3.   *A metric space X is embeddable in Hilbert space if and only if each of its finite subsets can be so embedded.*

Proof.   Only the sufficiency requires attention.

Choose a point $t_0 \in X$ and define $\mathscr{F}$ to be the collection of all finite subsets of X that contain t_0. Corresponding to each $F \in \mathscr{F}$ there exists a Hilbert space H_F and a map $\phi_F : F \to H_F$ such that $\phi_F(t_0) = 0$ and

$$\|\phi_F(s) - \phi_F(t)\| = \rho(s,t) \qquad (s,t \in F). \tag{2.6}$$

Corresponding to each $F \in \mathscr{F}$, let

$$S(F) = \{G : G \in \mathscr{F} \text{ and } G \supset F\}.$$

Each $S(F)$ is nonempty and

$$S(F_1) \cap S(F_2) \cap \cdots \cap S(F_n) = S(\bigcup_{j=1}^{n} F_j)$$

for each finite collection $F_1,F_2,\dots,F_n$ in $\mathscr{F}$. It follows from this observation that the collection of subsets of $\mathscr{F}$ defined by

$$\{G : G \subset \mathscr{F} \text{ and } G \supset S(F) \text{ for some } F \subset \mathscr{F}\}$$

is a filter in $\mathscr{F}$ [22]; hence by Zorn's lemma there exists an ultrafilter $\mathscr{U}$ in $\mathscr{F}$ such that

$$S(F) \in \mathscr{U} \text{ for all } F \in \mathscr{F}. \tag{2.7}$$

In the direct product $\prod_{F \in \mathscr{F}} H_F$, designate by H' that subspace consisting of all functions u on $\mathscr{F}$ such that $u(F) \in H_F$ $(F \in \mathscr{F})$ and such that the numbers $\|u(F)\|$ are bounded as F ranges over $\mathscr{F}$. (We are using $\|\cdot\|$ and $(\cdot,\cdot)$ for the norm and inner product in any one of the Hilbert spaces H_F.) On this subspace define an inner product by

$$(u,v) = \int_{\mathscr{F}} (u(F), v(F)) \, d\sigma(F) \qquad (u,v \in H'),$$

where σ is the finitely additive measure on the subsets of $\mathscr{F}$ defined by

$$\sigma(A) = \begin{cases} 1 & \text{if } A \in \mathcal{U} \\ 0 & \text{if } A \notin \mathcal{U}. \end{cases} \tag{2.8}$$

This last assertion is easily verified since the ultrafilter $\mathcal{U}$ contains precisely one of A and $\mathcal{F} \backslash A$ for every nonempty subset A of $\mathcal{F}$. The indicated integral is that of a bounded function with respect to a finitely additive measure on the algebra of all subsets of $\mathcal{F}$; hence it may be taken in the sense of successive refinements or in the context of integration with respect to a finitely additive set function as developed in Dunford-Schwartz [22]. Aside from the obvious properties we only need the following idea: a function $f : \mathcal{F} \to \mathbb{C}$ is σ-null or a *null function* provided that the set $\{F \in \mathcal{F} : |f(F)| > r\}$ has σ-measure 0 for each $r > 0$.

A σ-null function need not vanish almost everywhere with respect to σ; however, it is easy to verify that a function f is σ-null if and only if

$$\int_{\mathcal{F}} |f| \, d\sigma = 0.$$

It is clear that the statement $\|u - v\|$ is σ-null is an equivalence relation on H' which, in the usual fashion, permits a division of H' into mutually exclusive equivalence classes. We denote these equivalence classes by H, it being understood that all operations are taken modulo a σ-null function as it is done in the standard Lebesgue spaces. Relative to the norm

$$\|u\| = \left(\int_{\mathcal{F}} \|u(F)\|^2 \, d\sigma(F) \right)^{1/2} \qquad (u \in H),$$

H becomes an inner product space.

We are now in a position to define an isometric map ϕ from X into H: send the point $t \in X$ to the function $\phi(t)$ on $\mathcal{F}$ given by

$$\phi(t)(F) = \begin{cases} \phi_F(t) & \text{if } t \in F \\ 0 & \text{if } t \notin F. \end{cases}$$

For every F that contains t, $\|\phi(t)(F)\| = \|\phi_F(t)\| = \|\phi_F(t) - \phi_F(t_0)\| = \rho(t, t_0)$ so $\phi(t) \in H$. For each pair s, t of points in X let $A_{st} = S(\{s, t, t_0\})$. Then $A_{st} \in \mathcal{U}$ by (2.7) and $\sigma(A_{st}) = 1$ by (2.8). Also $\sigma(\mathcal{F} \backslash A_{st}) = 0$ since $\mathcal{F} \backslash A_{st} \notin \mathcal{U}$. Hence the norm of $\phi(s) - \phi(t)$ in H is, by (2.6),

$$\begin{aligned} \|\phi(s) - \phi(t)\|^2 &= \int_{\mathcal{F}} \|\phi(s)(F) - \phi(t)(F)\|^2 \, d\sigma(F) \\ &= \int_{A_{st}} \|\phi(s)(F) - \phi(t)(F)\|^2 \, d\sigma(F) \\ &= \rho(s, t)^2. \end{aligned}$$

Hence ϕ is an isometry of X into H^*, the completion of H. $\square$

These two lemmas combine in an obvious manner. Before formalizing, we prefer to express condition (2.1) in the following equivalent but more symmetric form:

$$\begin{cases} \sum_{j,k=0}^{n} \rho(x_j, x_k)^2 \, \xi_j \xi_k \leq 0 & \text{for all} \\ \text{choices of real numbers } \xi_0, \xi_1, \xi_2, \dots, \xi_n \text{ such that} \\ \sum_{j=0}^{n} \xi_j = 0. \end{cases} \tag{2.9}$$

In order to establish the equivalence start with (2.1) and sum over each term to get

$$2\sum_{j=1}^{n} \xi_j \sum_{k=1}^{n} \rho(x_0,x_k)^2 \, \xi_k - \sum_{j,k=1}^{n} \rho(x_j,x_k)^2 \, \xi_j\xi_k \geq 0,$$

put $\xi_0 = -\sum_{j=1}^{n} \xi_j$ and regroup to obtain (2.9); these steps are certainly reversible.

This slight recasting, at first glance rather innocuous in appearance, will prove to be fundamental.

Theorem 2.4. *A metric space (X,ρ) can be isometrically embedded in Hilbert space if and only if (2.9) holds for every finite subset $\{x_0,x_1,\ldots,x_n\}$ $(n \geq 2)$ of X.*

When applied to normed linear spaces the inequalities (2.9) may be viewed as characterizing inner-product spaces.

Theorem 2.5. *A real normed linear space B is an inner-product space if and only if*

$$\sum_{j,k=1}^{n} \|x_j - x_k\|^2 \, \xi_j\xi_k \leq 0 \quad when \quad \sum_{1}^{n} \xi_j = 0 \tag{2.10}$$

for every choice of points $x_1,x_2,\ldots,x_n$ $(n \geq 3)$ in B and real numbers $\xi_1,\xi_2,\ldots,\xi_n$.

Proof. It is easy to verify that (2.10) holds in any inner-product space.

Let B be a real normed linear space in which (2.10) is satisfied. Choose $n = 4$, $x_1 = x$, $x_2 = y$, $x_3 = -y$, $x_4 = 0$ and $\xi_1 = 1 - 2a$ $(0 < a < \tfrac{1}{2})$, $\xi_2 = \xi_3 = a$, $\xi_4 = -1$. Then (2.10) takes the form

$$(1-2a)\,a\|x-y\|^2 + (1-2a)\,a\|x+y\|^2 \leq (1-2a)\|x\|^2 + 2a(1-2a)\|y\|^2,$$

or $\quad \|x-y\|^2 + \|x+y\|^2 \leq a^{-1}\|x\|^2 + 2\|y\|^2.$

Letting $a\uparrow 1/2$, we obtain

$$\|x-y\|^2 + \|x+y\|^2 \leq 2[\|x\|^2 + \|y\|^2].$$

The reverse inequality follows by symmetry. Hence the norm in B satisfies the von Neumann-Jordan condition and so, B is an inner-product space [22, p.393]. $\square$

We turn now to some examples.

Example 2.6. Let ℓ_n^p, $1 \leq p \leq \infty$, denote $\mathbb{R}^n$ with the norm $\|x\|_p = \|(x_1,x_2,\ldots, x_n)\|_p = (\sum_1^n |x_j|^p)^{1/p}$ for $1 \leq p < \infty$ and $\|x\|_\infty = \max\{|x_j| : 1 \leq j \leq n\}$ when $p = \infty$. It follows directly from (2.1) that ℓ_n^p cannot be embedded in any Euclidean space when $n \geq 2$, $1 \leq p \leq \infty$ and $p \neq 2$.

It suffices to take $n = 2$. Make the restriction $1 \leq p < 2$, choose four points $x_0 = (1,1)$, $x_1 = (-1,1)$, $x_2 = (-1,-1)$ and $x_3 = (1,-1)$, and compute the matrix $A = (a_{jk})$ according to (2.4). The result is

$$A = \begin{pmatrix} 4 & \cdot 2^{1+2/p} & 4-2^{1+2/p} \\ 2^{1+2/p} & 2^{2+2/p} & 2^{1+2/p} \\ 4-2^{1+2/p} & 2^{1+2/p} & 4 \end{pmatrix}$$

With the choices $\xi_1 = \xi_3 = 1$ and $\xi_2 = -1$, we have

$$\sum_{j,k=1}^{3} a_{jk}\, \xi_j\, \xi_k = 2^3(2 - 2^{2/p}) < 0.$$

This inequality violates (2.1) thereby showing that embedding is impossible.

The impossibility of embedding ℓ_n^p in H when $n \geq 2$ and $2 < p \leq \infty$ is established in a similar manner by choosing the same ξ's with $x_0 = (1,0)$, $x_1 = (0,1)$, $x_2 = (-1,0)$ and $x_3 = (0,-1)$.

Example 2.7.   The proof of Theorem 2.1 shows that the process of embedding a finite metric space into Euclidean space rests upon the reduction of the quadratic form (2.1) to its canonical representation (2.5). As an illustration of this procedure consider the problem of constructing a regular n-simplex in $\mathbb{R}^n$ all of whose edges have length 1. Denoting the vertices of such a simplex by $A_0, A_1, \ldots, A_n$, the requirement is that

$$\|A_j - A_k\| = 1 \qquad (j \neq k, j,k = 0,1,\ldots,n)$$

and that the vertices not be in any $n-1$ dimensional hyperplane. This is equivalent to showing that a set of $n+1$ distinct points with the metric $\rho(s,t) = 1$ if $s \neq t$ and $\rho(s,s) = 0$ can be embedded in $\mathbb{R}^n$ but not $\mathbb{R}^{n-1}$. Relative to this metric the quadratic form (2.1) becomes

$$Q(\xi) = \sum_{j=1}^{n} \xi_j^2 + \sum_{j<k} \xi_j \xi_k.$$

From the identity

$$Q(\xi) = \sum_{j=1}^{n} \frac{j+1}{2j} (\xi_j + \frac{1}{j+1} \xi_{j+1} + \cdots + \frac{1}{j+1} \xi_n)^2,$$

we easily conclude that $Q(\xi)$ is positive definite and hence of rank n. Thus the embedding cannot lie in a proper subspace of $\mathbb{R}^n$ by Corollary 2.2. In fact, this last expression is the required canonical reduction and from it one may read off a choice of the vertices $A_k : A_0 = (0,0,\ldots,0)$, and the coordinates of A_k are

$$\left(\frac{1}{\sqrt{2 \cdot 2}}, \frac{1}{\sqrt{2 \cdot 2 \cdot 3}}, \frac{1}{\sqrt{2 \cdot 3 \cdot 4}}, \ldots, \frac{1}{\sqrt{2(k-1)k}}, \sqrt{\frac{k+1}{2k}}, \overbrace{0,0,\ldots,0}^{n-k} \right).$$

§3. Functions of Negative Type

Suppose X is a set, ρ is a real function on $X \times X$ and the map $\phi : X \to H$ satisfies (1.1). Then ρ is forced to have the following properties: For all points s, t and r in X,

  (a) $\rho(t,t) = 0$,
  (b) $\rho(s,t) = \rho(t,s) \geq 0$,
  (c) $\rho(s,t) \leq \rho(s,r) + \rho(r,t)$.
Thus ρ is a *semi-metric* on X. If, in addition,
  (d) $\rho(s,t) = 0$ implies $s = t$,
then ρ is *metric* for X.

The point of mentioning these obvious consequences of isometric embedding is to note that (d) is inessential. For example, consider the real line $\mathbb{R}$ equipped with the distance function

$$\rho(s,t) = |\sin(s-t)| \qquad (s \in \mathbb{R}, \ t \in \mathbb{R}).$$

Obviously ρ does not have property (d); however a little computation shows that

$$\phi(t) = (\tfrac{1}{2}\cos(2t), \tfrac{1}{2}\sin(2t)) \qquad (t \in \mathbb{R})$$

defines an embedding of the semi-metric space $(\mathbb{R},\rho)$ into $\mathbb{R}^2$.

A review of the proof of Theorem 2.1 reveals that property (d) is not involved, neither is (c); in fact the only metric properties invoked are (a) and (b), the triangular inequality (c) being a consequence of (2.1). In this way we are led to consider spaces whose distance functions satisfy only the properties (a) and (b).

Definition 3.1. *A quasi-metric space is a set X together with a real function ρ, called a quasi-metric, on $X \times X$ which has properties (a) and (b). A quasi-metric ρ is said to be of negative type if it satisfies the quadratic form condition (2.9) for every finite subset of X.*

Remark 3.2. It is important to note that the conclusion of Theorem 2.4 remains valid if, in the hypothesis, "quasi-metric" replaces "metric". Of course, a quasi-metric space which can be embedded in Hilbert space must at least be semi-metric, but such need not be the assumption ab initio.

Definition 3.3. *Let (X,ρ) be a quasi-metric space and F a nonnegative continuous function on $\mathbb{R}^+ = [0,\infty)$ such that $F(0) = 0$. The metric or F-transform of X is the set X with the quasi-metric*

$$\delta = F \circ \rho. \tag{3.1}$$

If δ is of negative type, we will say that F is a function of negative type on X. The class of functions of negative type on X will be denoted by $N(X)$.

For a given quasi-metric space X it is natural to ask for an intrinsic characterization of the class $N(X)$. The problem thus posed is equivalent to finding all functions F such that the F-transform of X can be embedded in Hilbert space. The following examples illustrate the concept of metric transform and functions of negative type.

Example 3.4. Consider the real line with the usual metric $|s-t|$ transformed by $F(t) = (t^2 + \sin^2(t))^{1/2}$, $0 \le t < \infty$, so that

$$\delta(s,t) = F(|s-t|) = ((s-t)^2 + \sin^2(s-t))^{1/2}.$$

The metric space $(\mathbb{R},\delta)$ is embedded in $\mathbb{R}^3$ by the map

$$\phi(t) = (t, 2^{-1}\cos(2t), 2^{-1}\sin(2t)) \qquad (t \in \mathbb{R}).$$

This is easily checked by verifying the identity

$$\begin{aligned}
\|\phi(s) - \phi(t)\|^2 &= (s-t)^2 + 4^{-1}(\cos(2s) - \cos(2t))^2 + 4^{-1}(\sin(2s) - \sin(2t))^2 \\
&= (s-t)^2 + \sin^2(s-t) = F(|s-t|)^2.
\end{aligned}$$

In the same setting, a choice of $F(t) = t|\cos(t)|$, $0 \le t < \infty$, gives the metric transform $\delta(s,t) = |s-t|\,|\cos(s-t)|$, but embedding is not possible because, contrary to (2.9),

$$\sum_{j,k=1}^{4} |x_j - x_k|^2 |\cos(x_j - x_k)|^2 \, \xi_j \xi_k > 0$$

for the choices $x_1 = 0, x_2 = \pi/4, x_3 = \pi/2, x_4 = 3\pi/4, \xi_1 = \xi_4 = 1$ and $\xi_3 = \xi_2 = -1$. Naturally the impossibility of embedding can also be deduced by noting that δ does not satisfy the triangular inequality.

Thus $(t^2 + \sin^2 t)^{1/2}$ is of negative type on $\mathbb{R}$ while $t|\cos t|$ is not.

Example 3.5.   Let $F(t) = |t|^\alpha$ $(0 < \alpha < 1)$ and remetrize $\mathbb{R}$ by

$$\delta(s,t) = |s-t|^\alpha. \tag{3.2}$$

In this somewhat more general case we rely on the integral formula

$$\begin{cases} |t|^{2\alpha} = c_\alpha^{-1} \int_0^\infty u^{-1-2\alpha} \sin^2(tu)du \\ c_\alpha = \int_0^\infty u^{-1-2\alpha} \sin^2 u\, du \end{cases} \qquad (0 < \alpha < 1) \tag{3.3}$$

to show that the metric space $(\mathbb{R},\delta)$ can be embedded in H.

Fix a positive integer n and choose real numbers $x_1, x_2, \dots, x_n$, $\xi_1, \xi_2, \dots, \xi_n$ with $\Sigma_1^n \xi_j = 0$. Using the metric (3.2), it follows from (2.9) that the inequality to be verified is

$$\sum_{j,k=1}^n |x_j - x_k|^{2\alpha} \xi_j \xi_k \leq 0.$$

We make the obvious substitution from (3.3) and utilize the identity

$$\sin^2(x_j - x_k)u = \sin^2(x_j u) + \sin^2(x_k u) - 2\sin^2(x_j u)\sin^2(x_k u) \\ - 2^{-1}\sin(2x_j u)\sin(2x_k u) \tag{3.4}$$

to obtain

$$\sum_{j,k=1}^n |x_j - x_k|^{2\alpha} \xi_j \xi_k = -c_\alpha^{-1} \int_0^\infty 2\left[\left(\sum_{j=1}^n \xi_j \sin^2(x_j u)\right)^2 \right. \\ \left. + 2^{-1}\left(\sum_{j=1}^n \xi_j \sin(2x_j u)\right)^2\right] u^{-1-2\alpha}\, du,$$

which is nonpositive. Further, assuming the x_j are distinct and nonzero, the vanishing of the integeral implies that $\Sigma_{j=1}^n \xi_j \sin(2x_j u) = 0$ for $0 \leq u < \infty$ and hence that $\|\xi\| = \|(\xi_1, \xi_2, \dots, \xi_n)\| = 0$. Consequently, the quadratic form (2.1) is positive definite and, therefore, the range of an embedding of $(\mathbb{R},\delta)$ into H cannot lie in a finite dimensional subspace by Corollary 2.2.

This special embedding problem, originally investigated by W. A. Wilson [91] in the case $\alpha = 1/2$, appears in [80] and motivates [66]. Also see §9.

§4. Radial Positive Definite Functions

Our principal goal in the study of functions of negative type on quasi-metric spaces is the characterization of the classes $N(H)$, $N(\mathbb{R}^N)$ and $N(L^p)$, $0 < p \leq \infty$. This will come somewhat later in §8-9. Our first step is a reformulation of the embedding criterion (2.1) in terms of a notion of positive definiteness.

Let E be a linear space over the real or complex field. A complex function on E is *positive definite* provided that

$$\sum_{j,k=1}^{n} f(x_j - x_k)\, \xi_j\, \bar{\xi}_k \geq 0 \tag{4.1}$$

for all choices of n points $x_1, x_2, \ldots, x_n$ ($n \geq 1$) in E and all complex numbers $\xi_1, \xi_2, \ldots, \xi_n$. We define $PD(E)$ to be the class of all such functions on E.

A positive definite function on E satisfies the relations

$$f(-x) = \overline{f(x)}, \tag{4.2}$$

$$|f(x)| \leq f(0), \tag{4.3}$$

and

$$|f(x) - f(0)|^2 \leq 2\, f(0)\, \mathrm{Re}[f(x) - f(0)] \tag{4.4}$$

for all $x \in E$. For a proof of these properties as well as the following theorem of Bochner, see [74] or [1].

Theorem 4.1. *A continuous complex function f on $\mathbb{R}^N$, $1 \leq N < \infty$, is positive definite if and only if there exists a finite and positive (≥ 0) Borel measure μ on R^N such that*

$$f(t) = \int_{R^N} e^{i(t,u)}\, d\mu(u) \qquad (t \in \mathbb{R}^N). \tag{4.5}$$

The link between isometric embedding and positive definite functions is furnished by the classical formula

$$\exp(-\lambda^2 t^2) = (2\sqrt{\pi})^{-1} \int_{-\infty}^{\infty} e^{i\lambda t u} \exp(-u^2/4)\, du, \tag{4.6}$$

which shows that $\exp(-\lambda^2 t^2)$ is positive definite on $\mathbb{R}$ for all real λ. To obtain the analogue in $\mathbb{R}^N$, replace t by t_j and u by u_j in (4.6) and multiply the resulting integrals to obtain

$$\exp(-\lambda^2 \|t\|^2) = (4\pi)^{-N/2} \int_{R^N} e^{i\lambda(t,u)} \exp(-\|u\|^2/4)\, du. \tag{4.7}$$

Here $du = du_1 du_2 \ldots du_N$, $(t,u) = t_1 u_1 + \ldots + t_N u_N$ and $\|t\|$ denotes the norm in $\mathbb{R}^N$. Thus the function $\exp(-\lambda \|t\|^2)$ is positive definite on $\mathbb{R}^N$ for $\lambda > 0$ and $N = 1, 2, \ldots$, hence positive definite on each finite-dimensional subspace of H. We summarize the result just proved:

Theorem 4.2. *The function $\exp(-\lambda \|t\|^2)$, $\lambda > 0$, is a positive definite function on Hilbert space H.*

In order to exploit this fact we introduce a variant of the standard notion of a positive definite function as given in (4.1). We will use $\mathbb{R}^+$ to denote the non-negative reals $[0, \infty)$.

Definition 4.3. *A real function F defined on $\mathbb{R}^+$ is radial positive definite on the quasi-metric space X provided that F is continuous and*

$$\sum_{j,k=1}^{n} F(\rho(x_j, x_k))\, \xi_j \xi_k \geq 0 \tag{4.8}$$

for all choices of n points $x_1, x_2, \ldots, x_n$ ($n \geq 1$) in X and real numbers $\xi_1, \xi_2, \ldots, \xi_n$. We denote the set of all such functions by $RPD(X)$.

The term "radial" is most descriptive in case X is a normed linear space, for then $F(\|x\|) = F(\|x-0\|)$. In view of this definition the content of Theorem 4.2 is that the function $\exp(-\lambda t^2)$ $(\lambda > 0)$ is positive definite on Hilbert space.

Theorem 4.4.   *The following hold in any quasi-metric space X.*
(a) $\mathrm{RPD}(X)$ *is never empty.*
(b) *If* $F_j \in \mathrm{RPD}(X), j=1,2$, *then* $F_1 \cdot F_2 \in \mathrm{RPD}(X)$.
(c) *If* $F_j \in \mathrm{RPD}(X)$ *and* $0 \le c_j < \infty, j=1,2,\ldots,n$, *then* $\Sigma c_j F_j \in \mathrm{RPD}(X)$.
(d) *If* $F_j \in \mathrm{RPD}(X), j=1,2,\ldots$ *and the* F_j *converge pointwise to a continuous limit* F, *then* $F \in \mathrm{RPD}(X)$.

Proof.   (a) The constant function 1 is always in $\mathrm{RPD}(X)$. (b) This is a consequence of a theorem of I. Schur to the effect that if (a_{jk}) and (b_{jk}) are two real $n \times n$ matrices such that the associated quadratic forms $\Sigma a_{jk}\xi_j\xi_k$ and $\Sigma b_{jk}\xi_j\xi_k$ are positive semidefinite, then so is $\Sigma a_{jk}b_{jk}\xi_j\xi_k$. For a proof see [72, p.307]. (c) and (d) are clear.

Theorem 4.5.   *In a quasi-metric space (X,ρ) the following are equivalent:*
(a) ρ *is of negative type;*
(b) *the function* $\exp(-\lambda t^2)$ *belongs to* $\mathrm{RPD}(X)$ *for* $\lambda > 0$;
(c) (X,ρ) *is embeddable in* H.

Proof.   The equivalence of conditions (a) and (c) is a restatement of Theorem 2.4 in the context of Remark 3.2. Hence we need only work with (a) and (b). According to (4.8) the implication (a)$\Rightarrow$(b) will follow once it is shown that

$$\sum_{j,k=1}^{n} \exp(-\lambda\rho^2(x_j,x_k))\xi_j\xi_k \ge 0 \qquad (4.9)$$

for every choice of points $x_1,x_2,\ldots,x_n$ in X and real ξ_j. Since ρ is of negative type there exists, by Theorem 2.1, points $y_1,y_2,\ldots,y_n$ in H such that $\|y_j-y_k\| = \rho(x_j,x_k)$ for $j,k=1,2,\ldots,n$. But these relations, when substituted in (4.9), result in the assertion that $\exp(-\lambda t^2)$, $\lambda > 0$, is a member of $\mathrm{RPD}(H)$, a fact already established by Theorem 4.2.

In order to verify that (b) implies (a), we start with (4.9) as given and expand the left side according to the exponential formula to obtain

$$\left|\sum_{1}^{n}\xi_j\right|^2 - \lambda\sum_{j,k=1}^{n} \rho^2(x_j,x_k)\xi_j\xi_k + \lambda^2\sum_{j,k=1}^{n} \rho^4(x_j,x_k)\tfrac{1}{2}\xi_j\xi_k - \cdots \ge 0.$$

Eliminate the first term by putting $\Sigma\xi_j = 0$, divide by positive λ and then let λ tend to 0; the limit is inequality (2.9). This proves (a). $\square$

The same argument establishes the fundamental relationship between the classes $N(X)$ and $\mathrm{RPD}(X)$.

Theorem 4.6.   *Suppose X is a quasi-metric space and F is a real-valued function on $\mathbb{R}^+$. The following are equivalent:*
(a) $F \in N(X)$.
(b) $\exp(-\lambda F^2) \in \mathrm{RPD}(X)$ *for each* $\lambda > 0$.
In particular, if X is a linear space and $F(\|x\|)$ is homogeneous, in the sense that

for some $\kappa > 0$, $F(\|\lambda x\|) = \lambda^\kappa F(\|x\|)$ *for* $x \in X$ *and* $\lambda > 0$, *then* (b) *may be replaced by the single condition*

(b') $\exp(-F^2) \in \mathrm{RPD}(X)$.

It is clear from the proof of Theorem 4.5 that (b) of Theorem 4.6 is equivalent to the requirement that $\exp(-\lambda_n \cdot F^2) \in \mathrm{RPD}(X)$ for some positive sequence λ_n with limit 0.

Also we should note that these last results further illuminate Example 3.4 where it was shown that $(t^2 + \sin^2 t)^{1/2}$ is of negative type on $\mathbb{R}$. By Theorem 4.6 it follows that $\exp(-\lambda(t^2 + \sin^2(t))) \in \mathrm{RPD}(\mathbb{R})$ for $\lambda > 0$ and hence is representable as a Fourier transform by Bochner's theorem. Such a representation is not possible for the functions $\exp(-\lambda t \cos^2(t))$ for any λ in some small neighborhood of zero, for otherwise we could apply the reasoning of Theorem 4.5 to conclude that $\delta(s,t) = |s - t| |\cos(s - t)|$ is a quasi-metric of negative type on $\mathbb{R}$, contrary to the example cited.

Theorem 4.7.   *If* $F \in N(X)$ *and* $0 < \alpha < 1$, *then*

$$F^\alpha \in N(X) \qquad\qquad (4.10)$$

and

$$\exp(-\lambda \, F^{2\alpha}) \in \mathrm{RPD}(X) \ \text{for} \ \lambda > 0. \qquad\qquad (4.11)$$

Proof.   This simple but useful result is a consequence of the integral formula

$$|t|^{2\alpha} = c_\alpha \int_0^\infty (1 - \exp(-\lambda^2 t^2))\lambda^{-1-2\alpha} \, d\lambda \quad (t \geq 0),$$

where

$$c_\alpha^{-1} = \int_0^\infty (1 - \exp(-\lambda^2))\lambda^{-1-2\alpha} \, d\lambda \quad (0 < \alpha < 1).$$

If F is a function of negative type on X, then $\exp(-\lambda F^2)$ belongs to $\mathrm{RPD}(X)$ for $\lambda > 0$. In order to show that F^α, $0 < \alpha < 1$ is also in $N(X)$, choose points $x_1, x_2, \ldots, x_n$ in X, real ξ_j with $\Sigma_1^n \, \xi_j = 0$ and make the obvious substitutions in (2.4) to get

$$\sum_{j,k=1}^n F^{2\alpha}(\rho(x_j,x_k))\xi_j\xi_k = c_\alpha \int_0^\infty \sum_{j,k=1}^n (1 - \exp(-\lambda^2 F^2(\rho(x_j,x_k))))\xi_j\xi_k \lambda^{-1-2\alpha} \, d\lambda$$

$$= -c_\alpha \int_0^\infty \sum_{j,k=1}^n \exp(-\lambda^2 F^2(\rho(x_j,x_k)))\xi_j\xi_k \lambda^{-1-2\alpha} \, d\lambda \leq 0.$$

This establishes (4.10). The membership assertion (4.11) now follows from the implication (a)$\Rightarrow$(b) of Theorem 4.6. $\square$

We mention the following immediate corollaries.

Corollary 4.8.   *Let* H_α *denote the metric space obtained from the Hilbert space H by changing the metric to* $\|x - y\|^\alpha$ $(0 < \alpha \leq 1)$. *Then*

(a) H_α *may be embedded in* H,

and

(b) H_γ *may be embedded in* H_α *for* $0 < \gamma \leq \alpha \leq 1$.

These observations are due to John von Neumann.

Proof. Since $\|x-y\|$ is of negative type on H, so is $\|x-y\|^\alpha$ by (4.10). Now apply Theorem 4.5. Obviously, (b) follows by applying (a) to $H_{\gamma/\alpha}$. $\square$

Corollary 4.9. *If F is homogeneous and $\exp(-F^2) \in \mathrm{RPD}(X)$, then $\exp(-F^{2\alpha})$ $\in \mathrm{RPD}(X)$ for $0 < \alpha < 1$. In particular $\exp(-\|x\|^{2\alpha})$ is positive definite on Hilbert space for $0 < \alpha \leq 1$.*

Notation. The space ℓ^p $(0 < p \leq \infty)$ denotes all complex number sequence $\{x_n\}_{n \geq 1}$ such that $\Sigma |x_n|^p < \infty$ for finite p and such that $\sup |x_n| < \infty$ when $p = \infty$. The distance between two points x and y in ℓ^p is given by

$$\|x-y\|_p = \left\{ \sum_1^\infty |x_n - y_n|^p \right\}^{1/p} \qquad (0 < p < \infty)$$

and

$$\|x-y\|_\infty = \sup_n |x_n - y_n| \qquad (p = \infty).$$

Notice that for p in the range $0 < p < 1$, the notion of distance is not a metric, but it is a quasi-metric.

Relative to a measure space (Ω, μ), $L^p(\mu)$ is defined to be the space of (equivalence classes of) complex μ-measurable functions on Ω having integrable pth powers for $0 < p < \infty$ and such that ess $\sup_\Omega |f| < \infty$ when $p = \infty$. The distance between two functions in $L^p(\mu)$ is defined by

$$\|f-g\|_p = \left\{ \int_\Omega |f-g|^p d\mu \right\}^{1/p} \qquad (0 < p < \infty)$$

and

$$\|f-g\|_\infty = \text{ess} \sup_\Omega |f-g| \qquad (p = \infty).$$

Theorem 4.10. *Let X denote any one of the spaces ℓ^p_m, ℓ^p or $L^p(\mu)$ with p in the range $0 < p \leq 2$. Then the following statements are true and equivalent by pairs: The function $\exp(-t^{2\alpha})$ is radial positive definite on X for $0 < \alpha \leq p/2$.* (4.12)
The function $F(t) = t^\alpha$ is a function of negative type on X for $0 < \alpha \leq p/2$. (4.13)
The space X with the metric $\|x-y\|_p^\alpha$ is embeddable in H for $0 < \alpha \leq p/2$. (4.14)

$$\sum_{j,k=1}^n \|x_j - x_k\|_p^{2\alpha} \, \xi_j \xi_k \leq 2 \sum_{j=1}^n \xi_j \|x_j\|_p^{2\alpha} \text{ for any choice of } n \qquad (4.15)$$

points $x_1, x_2, \ldots, x_n$ $(n \geq 2)$ in X and real numbers $\xi_1, \xi_2, \ldots, \xi_n$ satisfying $\Sigma_1^n \xi_j = 1$, with $0 < \alpha \leq p/2$.

Proof. First we consider (4.12) for the case $X = \ell^p$. Define the homogeneous coordinate maps p_j on ℓ^p by

$$p_j(x) = p_j((x_1, x_2, \ldots)) = |x_j| \qquad (1 \leq j < \infty).$$

Each of the p_j is a negative type semi-metric since the map $x \to (\mathrm{Re}\ x_j, \mathrm{Im}\ x_j)$ defines an embedding of the semi-metric space (ℓ^p, p_j) into $\mathbb{R}^2$; hence upon putting $2\alpha = p$ in (4.11) we conclude that $\exp(-|x_j|^p)$ is a radial positive definite function on ℓ^p. Moreover, the finite product of such functions remains in $\mathrm{RPD}(\ell^p)$ by Theorem 4.4(b). Thus the functions

$$\exp(-[|x_1|^p + |x_2|^p + \cdots + |x_j|^p]) \qquad (j=1,2,\ldots)$$

are positive definite on ℓ^p. It follows from this fact and a simple limiting argument in (4.8) that $\exp(-\|x\|_p^p)$ is positive definite on ℓ^p or, what is the same thing, $\exp(-t^p)$ is radial positive definite on ℓ^p. Thus, by Theorem 4.6, the function $t^{p/2}$ is of negative type on ℓ^p and hence, by Theorem 4.7, $t^* = t^{\delta p/2}$ is also of negative type on ℓ^p where $\delta = 2\alpha/p \le 1$. Hence $\exp(-t^{2\alpha}) \in \mathrm{RPD}(\ell^p)$ when $0 < \alpha \le p/2$ by (4.11).

The equivalence of (4.12) with (4.13) is Theorem 4.6 and the equivalence of (4.13) and (4.14) follows from Theorem 4.5. For the equivalence of (4.14) and (4.15) we note that, by Theorem 2.4, (4.14) is equivalent to the inequality

$$\sum_{j,k=1}^{n} (\|x_j - x_0\|_p^{2\alpha} + \|x_k - x_0\|_p^{2\alpha} - \|x_j - x_k\|_p^{2\alpha})\xi_j\xi_k \ge 0 \qquad (4.16)$$

for all choices of n points ($n \ge 2$) in ℓ^p and real numbers ξ_j. The inequality (4.15) results from setting $x_0 = 0$, requiring that $\Sigma_1^n \xi_j = 1$ and then summing over j and k separately. Conversely, if (4.15) holds it is also true for $\Sigma\xi_j = r \ne 0$, so also for $\Sigma\xi_j = 0$ and hence (4.16) holds. Thus (4.14) follows from Theorem 2.4. This completes the proof for ℓ^p and ℓ_n^p.

There is no difficulty in extending these statements to $L^p(\mu)$. Since the equivalences follow as before it is only necessary to show, for example, that $\exp(-t^p)$ is a radial positive definite function on L^p. Choose simple measurable functions $x_1, x_2, \ldots, x_n$ on Ω and let $A_1, A_2, \ldots, A_n$ be a pairwise disjoint collection of μ-measurable subsets of Ω such that $x_j = c_{ji}$ on A_i for $1 \le j \le n$, $1 \le i \le m$. It is required to show that

$$\sum_{j,k=1}^{n} \exp(-\|x_j - x_k\|_p^p)\xi_j\xi_k \ge 0 \qquad (4.17)$$

for real ξ_j. Since the integrals reduce to

$$\|x_j - x_k\|_p^p = \int_\Omega |x_j(\omega) - x_k(\omega)|^p d\mu(\omega) = \sum_{i=1}^{m} |c_{ji} - c_{ki}|^p \, \mu(A_i)$$

$$= \sum_{i=1}^{m} |\mu(A_i)^{1/p} c_{ji} - \mu(A_i)^{1/p} c_{ki}|^p,$$

the right-hand sum represents the pth power of the distance between two points in ℓ^p. Hence, using the corresponding result in ℓ^p, we see that (4.17) holds for simple measurable functions and consequently must hold in general. Now apply Corollary 4.9 and the previously established equivalences. $\square$

We shall have use for the quadratic inequality (4.15) in §19 where it has a primary role in the problem of extending Lipschitz-Hölder maps from L^p into H. Other related inequalities concerning L^p spaces are derived in §15.

Although not a metric space for $0 < p < 1$, the space L^p does become a metric space if its metric is raised to the power α for $0 < \alpha \le p$; and when α is further restricted to satisfy $0 < \alpha \le p/2$ then the resulting space is, by (4.14), embeddable in H. Stated another way, for α in the range $0 < \alpha \le p/2$ there exists a map $\phi : L^p \to H$ such that

$$\|\phi(x) - \phi(y)\| = \|x - y\|_p^\alpha \qquad (x \in L^p, \ y \in L^p).$$

The preceding results imply an important corollary.

Corollary 4.11. *The function* $\exp(-|t|^{2\alpha})$ *is not positive definite on* $\mathbb{R}$ *when* $\alpha > 1$ *and it is not positive definite on* ℓ^p *when* $\alpha > p/2$ *and* $0 < p \leq 2$.

Proof. If the indicated function were positive definite on $\mathbb{R}$ for some $\alpha > 1$, then $(\mathbb{R}, |x - y|^\alpha)$ would have to be embeddable in Hilbert space, but such is not the case since $|x - y|^\alpha$ $(\alpha > 1)$ is not even a metric on $\mathbb{R}$. The second assertion follows, for example, by noting that the analogue of the matrix A of Example 2.6, based on the points $x_0 = (1,1)$, $x_1 = (-1,1)$, $x_2 = (-1,-1)$, and $x_3 = (1,-1)$ and the metric $\|x - y\|_p^\alpha$ is not positive definite when $\alpha > p/2$. $\square$

§5. A Characterization of Subspaces of $L^p, 1 \leq p \leq 2$

The characterization of inner-product spaces provided by (2.10) suggests an investigation of those real normed linear spaces B which, for some p, satisfy the condition

$$\sum_{j,k=1}^n \|x_j - x_k\|^p \xi_j \xi_k \leq 0 \quad \text{when} \quad \sum_1^n \xi_j = 0$$

for all $x_1, x_2, \ldots, x_n$ in B and real ξ_j. This condition is nothing but the statement that $\|x - y\|^{p/2}$ is of negative type on B, which is known to be true (see (4.13)) when B is an L^p space and $0 < p < 2$. The fundamental result of this section, due to Bretagonolle, Dacunha Castelle, and Krivine, is that this is the only possibility when $1 \leq p \leq 2$ [11].

Theorem 5.1. *In order that a real normed linear space B be linearly isometric to a subspace of some L^p space, $1 \leq p \leq 2$, it is necessary and sufficient that $\|x - y\|^{p/2}$ be of negative type on B.*

As we have seen this result is true and not terribly difficult when $p = 2$. It is much more demanding for $1 \leq p \leq 2$ and follows only after a sequence of results which are themselves of independent interest. The first of these complements the work of S. Kakutani on *abstract (L)-spaces* [40].

Definition 5.2. *A Banach space B will be called an abstract (L^p)-space $(1 \leq p < \infty)$ provided that, first, it is a vector lattice. Thus there exists a relation $x \leq y$ between some pairs $x, y \in B$ which satisfies the following conditions for all $x, y, z, w \in B$ and $\lambda \in R$:*

$$x \geq y, \ y \geq x \ \text{imply} \ x = y, \tag{5.1}$$
$$x \geq y, \ y \geq z \ \text{imply} \ x \geq z, \tag{5.2}$$
$$x \geq y, \ \lambda \geq 0 \ \text{imply} \ \lambda x \geq \lambda y, \tag{5.3}$$
$$x \geq y \ \text{implies} \ x + z \geq y + z. \tag{5.4}$$

For each pair $x, y \in B$, there is a unique supremum $z = x \vee y$ $\qquad$ (5.5)
such that $z \geq x$, $z \geq y$ and if $z' \geq x$, and $z' \geq y$ then $z' \geq z$.
For each pair $x, y \in B$, there is a unique infimum $w = x \wedge y$ $\qquad$ (5.6)
such that $w \leq x$, $w \leq y$ and if $w' \leq x$ and $w' \leq y$ then $w' \leq w$.
Secondly, the partial ordering is required to be compatible with norm convergence in B in the sense that

$$x_n \geq y_n \text{ and } \|x_n - x\| \to 0, \|y_n - y\| \to 0 \text{ imply } x \geq y. \tag{5.7}$$

Finally, it is required that

$$\|\,|x|\,\| = \|x\|, \tag{5.8}$$

$$x \geq 0, \; y \geq 0 \text{ imply that } \|x+y\|^p \geq \|x\|^p + \|y\|^p \geq \|x \vee y\|^p. \tag{5.9}$$

Each element x in B has a *positive part, a negative part* and a *modulus* defined, respectively, by

$$x^+ = x \vee 0, \; x^- = (-x) \vee 0, \text{ and } |x| = x \vee (-x).$$

Two elements x and y are called *disjoint* provided that $|x| \wedge |y| = 0$. We shall now state some fundamental facts about linear lattices which will be needed in the following presentation.

$$\lambda \geq 0 \text{ implies } \lambda(x \vee y) = (\lambda x) \vee (\lambda y), \; \lambda(x \wedge y) = (\lambda x) \wedge (\lambda y). \tag{5.10}$$

$$(x \vee y) + z = (x+z) \vee (y+z), \; (x \wedge y) + z = (x+z) \wedge (y+z). \tag{5.11}$$

$$x + y = x \vee y + x \wedge y. \tag{5.12}$$

$$x = x^+ - x^-, \; x^+ \wedge x^- = 0. \tag{5.13}$$

$$(x \vee y) \wedge z = (x \wedge z) \vee (y \wedge z), \; (x \wedge y) \vee z = (x \vee z) \wedge (y \vee z). \tag{5.14}$$

$$|x+y| \leq |x| + |y|. \tag{5.15}$$

$$2(x \vee y) = (x+y) + |x-y|. \tag{5.16}$$

$$2(x \wedge y) = x + y - |x-y|.$$

$$\big|\,|x| - |y|\,\big| \leq |x-y|. \tag{5.17}$$

$$|x_i| \wedge |x_j| = 0 \; (i \neq j, \; i,j = 1,2,\dots,n) \text{ implies} \tag{5.18}$$

$$|x_1 + x_2 + \cdots + x_n| = |x_1| + |x_2| + \cdots + |x_n| = |x_1| \vee |x_2| \vee \cdots \vee |x_n|.$$

Proofs of these properties and the general theory of linear lattices will be found in G. Jameson [39].

Theorem 5.3. *Fix $1 \leq p < \infty$ and let B be an abstract (L^p)-space. Then there exists a measure space (Ω, μ) such that B is isometric and lattice-isomorphic to $L^p(\Omega, \mu)$.*

Proof. Let us first list some elementary lemmas that hold in any abstract (L^p)-space.

(1) If $x \wedge y = 0$, then $\|x \vee y\|^p = \|x+y\|^p = \|x\|^p + \|y\|^p$.

(2) If $x_1, x_2, \dots, x_n$ are disjoint by pairs, then

$$\|x_1 + x_2 + \cdots + x_n\|^p = \|x_1\|^p + \|x_2\|^p + \cdots + \|x_n\|^p.$$

(3) If $0 \leq x \leq y$, then $\|x\| \leq \|y\|$.

(4) If $x_n \leq x_{n+1} \leq x$ for $n = 1,2,\dots$, then the sequence x_n has a limit.

(5) The functions $x \vee y$ and $x \wedge y$ are continuous from $B \times B$ to B.

For (1), note that if $x \wedge y = 0$, then, necessarily, $x, y \geq 0$ so, by (5.9) and (5.12), $\|x+y\|^p \geq \|x\|^p + \|y\|^p \geq \|x \vee y\|^p = \|x+y\|^p$.

The assertion (2) follows from (1), (5.18) and a simple induction argument.

The assumption $0 \leq x \leq y$ in (3) implies that $y = x + a$ with $a \geq 0$. Hence, by (5.9), $\|y\|^p = \|x+a\|^p \geq \|x\|^p + \|a\|^p \geq \|x\|^p$ and, taking pth roots, $\|x\| \leq \|y\|$.

For (4) it is clearly enough to show that every sequence x_n in B such that $x_n \geq x_{n+1} \geq 0 \; (n = 1,2,\dots)$ has a limit. By (3), $\|x_n\|$ is a decreasing number sequence and thus has a limit. If $n > m$, $\|x_n - x_m\|^p \leq \|x_n\|^p - \|x_m\|^p$ by (5.9) so that the sequence x_n is Cauchy.

The assertion of continuity in (5) is an immediate consequence of (5.8), (5.15), (5.16) and (5.17).

Corresponding to each e in B with $e \geq 0$ we define a subspace $M(e)$ by

$$M(e) = \{x \in B : ne \geq |x| \text{ for some } n \in \mathbb{R}^+\}.$$

It is clear that $M(e)$ is also a vector lattice. Furthermore, if the elements $e_1, e_2, \ldots, e_k$ in B are ≥ 0 and disjoint by pairs, then the subspaces $M(e_1), M(e_2), \ldots, M(e_k)$ are linearly independent. For otherwise there would exist elements $u_j \in M(e_j)$, $1 \leq j \leq k$, and real numbers $\lambda_1, \ldots, \lambda_k$ with $\lambda_1 u_1 + \lambda_2 u_2 + \cdots + \lambda_k u_k = 0$ with, say, $\lambda_1 \neq 0$ and $u_1 \neq 0$. Thus $u_1 = \alpha_2 u_2 + \cdots + \alpha_k u_k$ and

$$|u_1| \leq |\alpha_2|\,|u_2| + \cdots + |\alpha_k|\,|u_k|.$$

Since there exist $n_j \in R^+$ such that $|u_j| \leq n_j e_j$ for $1 \leq j \leq k$, we have

$$|u_1| \leq M(e_2 + e_3 + \cdots + e_k) = M(e_2 \vee e_3 \vee \ldots \vee e_k)$$

when $M = \max(n_2|\alpha_2|, \ldots, n_k|\alpha_k|)$. But $|u_1| \leq n_1 e_1$ so, with $N = \max(M, n_1)$, these combine to give

$$|u_1| \leq N\, e_1 \wedge (e_2 \vee \ldots \vee e_k) = 0.$$

Whence $u_1 = 0$, contrary to hypothesis.

Now let J denote a subset of B which is maximal with respect to the property of consisting of elements of norm 1 which are ≥ 0 and disjoint by pairs; the existence of such a set is guaranteed by Zorn's lemma. Let M denote the smallest subspace of B which contains all the subspaces $M(e)$ for $e \in J$. According to the preceding, M is just the direct sum of the subspaces $M(e)$ $(e \in J)$.

We now show that M is dense in B. Since every element decomposes into the difference of two positive elements, it will suffice to show that if $x \geq 0$ then $x \in \overline{M}$. It is certainly true that

$$x \geq \bigvee_{e \in I}(x \wedge e) = \sum_{e \in I}(x \wedge e) \geq 0$$

for every finite subset I of J. By (2) and (3)

$$\|x\|^p \geq \sum_{e \in I} \|x \wedge e\|^p;$$

hence the set $\{e \in J : x \wedge e \neq 0\}$ is a countable set $\{e_1, e_2, \ldots\}$. Let us now put

$$f = \sum_{k=1}^{\infty} e_k/2^k$$

and

$$x_n = x \wedge (nf) \qquad (n = 1, 2, 3, \ldots).$$

The series converges because $\|e_k\| = 1$. By (5) we may write

$$x_n = \lim_{N \to \infty} \sum_{k=1}^{N} x \wedge (ne_k/2^k) = \sum_{k=1}^{\infty} x \wedge (ne_k/2^k).$$

The sequence x_n is nondecreasing and satisfies $x_n \leq x$; hence it has a limit u by (4), and $u \in M$. In fact $u = x$. To see this put $v = x - u$ and $w = v \wedge f$. Clearly $w \leq v \leq x - x \wedge (nf)$ for $n = 1, 2, \ldots$. Evidently $w \leq x$ and, supposing that $kw \leq x$, we have from

$w + x \wedge (kf) \le x$ that $w + (kw) \wedge (kf) \le x$ or $w + k(w \wedge f) \le x$. But $w \le f$, hence $w = w \wedge f$ and $(1 + k)w \le x$. It follows by induction that $kw \le x$ for all $k = 1, 2, \ldots$. Since $w \ge 0$ this implies that $w = 0$.

From $v \wedge f = 0$ it follows that $v \wedge e_n = 0$ for all n and from $0 \le v \le x$ it follows that $v \wedge e = 0$ for $e \in J \backslash \{e_1, e_2, \ldots\}$. Hence $v = 0$ for otherwise the maximality of J would be contradicted by the set $J \cup \{v / \|v\|\}$.

The next major step is to show that if $e \in J$, then $\overline{M(e)}$ is isomorphic to $L^p(\Omega_e, \mu_e)$ for some measure space (Ω_e, μ_e). We define

$$B(e) = \{u \in M(e) : u \wedge (e - u) = 0\}.$$

Clearly every $u \in B(e)$ satisfies $0 \le u \le e$. Further, if $u, v \in B(e)$, then

$$e - u \vee v = -(u - e) \vee (v - e) = (e - u) \wedge (e - v)$$

and

$$\begin{aligned}(u \vee v) \wedge (e - u \vee v) &= (u \vee v) \wedge [(e - u) \wedge (e - v)] \\ &= [u \wedge (e - u) \wedge (e - v)] \vee [v \wedge (e - u) \wedge (e - v)] = 0.\end{aligned}$$

Hence $u \vee v \in B(e)$.

If u_n is a nondecreasing sequence of elements in $B(e)$ then, since $u_n \le e$, u_n has a limit u and $u \in B(e)$ because, by (5),

$$u \wedge (e - u) = \lim_{n \to \infty} u_n \wedge (e - u_n) = 0.$$

Finally, each $u \in B(e)$ has a complement $u' = e - u$ such that $u \vee u' = e$.

These remarks show that $B(e)$ is a Boolean σ-algebra. Hence $B(e)$ has a concrete representation as the Boolean algebra K_e of all the simultaneously open and closed subsets of some totally disconnected compact Hausdorff space Ω_e [34]. If A_u is the element of Ω_e corresponding to the element u in $B(e)$, then $\mu_e(A_u) = \|u\|^p$ clearly (see (1)) defines a finitely additive measure on K_e. Further μ_e can be extended to the σ-algebra $\mathscr{A}$ on Ω_e generated by K_e, and $\mathscr{A}$, modulo the σ-ideal $\mathscr{A}_0$ of sets of μ_e-measure zero in $\mathscr{A}$, forms a Boolean algebra which is lattice-isomorphic to $B(e)$.

Let $S(e)$ denote the subspace of $M(e)$ formed by the real linear combinations of the elements of $B(e)$. Corresponding to each $u \in B(e)$, let $f(u)$ denote the characteristic function of the set A_u. The map

$$\sum_{j=1}^n c_j u_j \to \sum_{j=1}^n c_j f(u_j) \quad (c_1, c_2, \ldots, c_n \in R; \; u_1, u_2, \ldots, u_n \in B(e))$$

defines an isomorphism from $S(e)$ to the real simple measurable functions on Ω_e, modulo the subspace of functions zero almost everywhere relative to μ_e. Since the isomorphism is an isometry in the $L^p(\Omega_e, \mu_e)$ norm it extends to an isometry and lattice-isomorphism from $\overline{S(e)}$ to $L^p(\Omega_e, \mu_e)$. In order to show that $S(e)$ is dense in $M(e)$ and hence that $\overline{S(e)} = \overline{M(e)}$, we need the following fact:

> If $x \in M(e)$, $x \ge 0$ and $x \ne 0$, then there exists $v \in B(e)$,   (5.19)
> $v \ne 0$ and an integer $n > 0$ such that $x \ge \frac{1}{n} v$.

After multiplication by some positive integer we may suppose that $x \not\le e$ for if $kx \le e$ for all k then $x = 0$ contrary to assumption. First put

$$z = (x - e) \vee 0 \quad \text{and} \quad v = \lim_{n \to \infty} e \wedge nz.$$

Since $z > 0$ and $z \in M(e)$, then $z \wedge e > 0$ so that $v > 0$ since $v > z \wedge e$. For each positive integer, let $w_k = (e - v) \wedge kz$. Then $w_k \leq e - v \leq e - e \wedge nz$ for all $n > 0$ and $0 \leq w_k \leq kz$. It is certainly true that $w_k \leq e$ and if $nw_k \leq e$ for some positive integer n, then, since $w_k \leq e - e \wedge pz$ holds for all $p > 0$, $w_k \leq e - e \wedge nkz$ and thus $w_k + e \wedge nkz \leq e$ or $w_k + n(w_k \wedge kz) \leq e$. But $w_k \leq kz$ so $w_k \wedge kz = w_k$ and $(1 + n)w_k \leq e$. It follows by induction that $0 \leq nw_k \leq e$ for all $n = 1, 2, \ldots$ and hence that $w_k = 0$.

We now have that $(e - v) \wedge kz = 0$ and also that $(e - v) \wedge e \wedge kz = 0$ for $k = 1, 2, \cdots$. From (5) and the definition of v, it follows that

$$0 = \lim_{k \to \infty} (e - v) \wedge (e \wedge kz) = (e - v) \wedge v.$$

This shows that $v \in B(e)$. Further, if $w = e \wedge n(x - e)$ one has $w \leq e$, $w \leq nx - ne$ so $w + ne \leq nx$ or $(n + 1)w \leq nx$. Hence $x \geq (1 + \frac{1}{n})w$ and, as $x \geq 0$, this implies

$$x \geq (1 + \tfrac{1}{n})(w \vee 0) \geq w \vee 0.$$

Hence

$$x \geq [e \wedge n(x - e)] \vee 0 = e \wedge [n(x - e) \vee 0] = e \wedge nz,$$

so, letting $n \to \infty$, we see that $x \geq v$ as required.

We can now show that $S(e)$ is dense in $M(e)$. Suppose the conclusion false. Then there exists an $x \in M(e)$ such that $x \geq 0$ and $x \notin \overline{S(e)}$. Put $m = \sup\{\|y\| : y \in S(e)$ and $0 \leq y \leq x\}$ and choose a sequence y_n of these y's such that $\|y_n\| \to m$. Through the device of replacing each y_n by $\bigvee_{j=1}^{n} y_j$ we may assume that the y_n are non-decreasing. Then $\lim_{n \to \infty} y_n = y$, $\|y\| = m$ and $0 \leq y \leq x$. As $x - y > 0$, it follows from (5.19) that there exists a $v \in B(e)$, $v \neq 0$, and an integer $n > 0$ such that $x - y \geq \frac{1}{n}v$. Then $y + \frac{1}{n}v \in \overline{S(e)}$, $0 \leq y + \frac{1}{n}v \leq x$ and

$$\|y + \tfrac{1}{n}v\|^p \geq \|y\|^p + \|\tfrac{1}{n}v\|^p > \|y\|^p.$$

Hence $\|y + \frac{1}{n}v\| > m$ contrary to the definition of m. This proves that $\overline{S(e)} = \overline{M(e)}$.

In summary, we have shown that for each $e \in J$, there exists a measure space (Ω_e, μ_e) such that $\overline{M(e)}$ is isometric and lattice-isomorphic to $L^p(\Omega_e, \mu_e)$. As e ranges over J, the subspaces $M(e)$ are linearly independent by pairs and their direct sum over J defines a dense subspace M of B. Now let (Ω, μ) be the measure space determined by the disjoint union of the measures spaces (Ω_e, μ_e) $(e \in J)$. In the obvious way we define an isomorphism ϕ from M into $L^p(\Omega, \mu)$, and ϕ is certainly an isometry because if $x_j \in M(e_j)$, $j = 1, 2, \ldots, n$, then $|x_1|, |x_2|, \ldots, |x_n|$ are disjoint by pairs and

$$\|x_1 + x_2 + \cdots + x_n\|^p = \|x_1\|^p + \|x_2\|^p + \cdots + \|x_n\|^p.$$

Since $\phi(M)$ is dense in $L^p(\Omega, \mu)$, it extends in a unique way to an isometric isomorphism from B onto $L^p(\Omega, \mu)$. $\square$

Definition 5.4. *A real normed linear (metric) space is of type p $(0 < p < \infty)$ provided it is linearly isometric (isometric) to a subspace (subset) of some L^p space.*

Theorem 5.5. *A real normed linear space B is of type p $(1 \leq p < \infty)$ if and only if each of its finite-dimensional subspaces is of type p.*

Proof. The condition is clearly necessary.

For the sufficiency suppose that corresponding to every finite-dimensional subspace F of B there is a measure space (Ω_F, μ_F) and a linear isometry ϕ_F from F into $L_F^p = L^p(\Omega_F, \mu_F)$.

Let $\mathscr{F}$ denote the collection of finite-dimensional subspaces of B and for each $F \in \mathscr{F}$ put

$$S(F) = \{G : G \in \mathscr{F} \text{ and } G \supset F\}.$$

Exactly as in the proof of Lemma 2.3 there exists an ultrafilter $\mathscr{U}$ in $\mathscr{F}$ such that

$$S(F) \in \mathscr{U} \quad \text{for all} \quad F \in \mathscr{F}. \tag{5.20}$$

And—still following that proof—let σ denote the finitely additive measure on the collection of all subsets of $\mathscr{F}$ such that $\sigma(A) = 1$ if $A \in \mathscr{U}$ and $\sigma(A) = 0$ otherwise. From the product of the spaces L_F^p ($F \in \mathscr{F}$), whose points we denote by

$$\{f_F\}_{F \in \mathscr{F}} \quad \text{with} \quad f_F \in L_F^p, \tag{5.21}$$

let us single out the subspace L consisting of those equivalence classes (relative to the notion of σ-null function) of elements for which the numbers $\|f_F\| = $ norm of f_F in L_F^p remain bounded as F varies over $\mathscr{F}$. Now define a norm in L by the formula

$$\|\{f_F\}\| = \left(\int_{\mathscr{F}} \|f_F\|^p \, d\sigma(F)\right)^{1/p}. \tag{5.22}$$

In L we define the lattice operations by

$$\{f_F\} \vee \{g_F\} = \{f_F \vee g_F\} \quad \text{and} \quad \{f_F\} \wedge \{g_F\} = \{f_F \wedge g_F\}, \tag{5.23}$$

and the induced partial order is

$$\{f_F\} \geq \{g_F\} \quad \text{if and only if} \quad \{f_F\} \wedge \{g_F\} = \{g_F\}. \tag{5.24}$$

If $\{f_F\}$, $\{f_F'\}$, $\{g_F\}$, $\{g_F'\}$ are elements of L then

$$\|f_F \vee g_F - f_F' \vee g_F'\| \leq \|f_F - f_F'\| + \|g_F - g_F'\|,$$

by (5.16), and hence

$$\|\{f_F\} \vee \{g_F\} - \{f_F'\} \vee \{g_F'\}\| \leq \|\{f_F\} - \{f_F'\}\| + \|\{g_F\} - \{g_F'\}\|$$

by (5.22). This inequality shows that the lattice operation (5.23) is well defined on the equivalence classes of L and that $\vee$ is a continuous map from $L \times L$ into L. Hence L satisfies the conditions (5.1) to (5.6) of a vector lattice plus condition (5.7). Further, its completion L' is an ordered Banach lattice.

Since $|\{f_F\}| = \{|f_F|\}$ it follows that $\|\,|\{f_F\}|\,\| = \|\{f_F\}\|$ and if $\{f_F\}$, $\{g_F\} \geq 0$ then

$$\|\{f_F\} + \{g_F\}\|^p = \|\{f_F \vee 0\} + \{g_F \vee 0\}\|^p \geq \|\{f_F \vee 0\}\|^p + \|\{g_F \vee 0\}\|^p$$

$$= \|\{f_F\}\|^p + \|\{g_F\}\|^p \geq \|\{f_F\} \vee \{g_F\}\|^p.$$

Together, these assertions and inequalities show that L' is an abstract (L^p)-space. Hence by Theorem 5.3 there exists a measure space (Ω,μ) such that L' is linearly isometric to $L^p(\Omega,\mu)$.

To complete the proof we define a map ϕ from B into L by

$$\phi(x) = \{f_F\}_{F\in\mathscr{F}} \quad \text{with} \qquad \begin{aligned} f_F &= \phi_F(x) \;\; \text{if} \;\; x\in F \\ f_F &= 0 \;\; \text{if} \;\; x\notin F. \end{aligned}$$

To see that ϕ is linear choose $x,y\in B$ and note that $\phi(x+y)(F) = \phi(x)(F) + \phi(y)(F)$ whenever both x and y belong to $F\in\mathscr{F}$. Since the set

$$A_{xy} = \{F\in\mathscr{F} : x,y,\in F\}\notin\mathscr{U}$$

by (5.20), it follows that $\sigma(A_{xy}) = 1$ and $\sigma(\mathscr{F}\setminus A_{xy}) = 0$. Hence

$$\|\phi(x+y)-\phi(x)-\phi(y)\| = \left(\int_{A_{xy}} \|\phi(x+y)(F)-\phi(x)(F)-\phi(y)(F)\|^p d\sigma(F)\right)^{1/p}$$

$$= 0.$$

In a similar way it follows that $\phi(\lambda x) = \lambda\phi(x)$ for all $\lambda\in R$ and $x\in B$.

Finally, since $\|\phi_F(x)\| = \|x\|$ when $x\in F$ and

$$A_x = \{F\in\mathscr{F} : x\in F\}\in\mathscr{U},$$

we have

$$\|\phi(x)\| = \left(\int_{A_x} \|\phi(x)(F)\|^p d\sigma(F)\right)^{1/p} = \|x\|,$$

so ϕ is an isometry. This completes the proof. $\square$

Minor alterations of the above proof lead to the following important corollary.

Corollary 5.6. *A metric space (X,ρ) can be isometrically embedded into some L^p space $(p\geq 1)$ if and only if each finite subset of X can be so embedded.*

Lemma 5.7. *Suppose B is a real normed linear space, $0<\alpha\leq 2$ and ρ is a map from B to $\mathbb{R}^+$ such that*
  (a) *$\rho(\lambda_1 t_1 + \lambda_2 t_2 + \cdots + \lambda_n t_n)$ is continuous in $(\lambda_1,\lambda_2,\dots,\lambda_n)\in\mathbb{R}^n$ for every choice of $t_1,t_2,\dots,t_n\in B$,*
  (b) *$\rho(\lambda t) = |\lambda|\,\rho(t)$ for all $\lambda\in\mathbb{R}$ and $t\in B$,*
  (c) *$\rho(t-s)^{\alpha/2}$ is of negative type on B.*
There exists a probability space (Ω,μ) and a map $\phi:t\to X_t$ from B into $L^\beta(\Omega,\mu)$ — for every $\beta\in(0,\alpha)$ if $\alpha<2$ and for all $\beta>0$ if $\alpha=2$ — such that

$$\rho(t) = \|\phi(t)\|_\beta = \left(\int_\Omega |X_t|^\beta d\mu\right)^{1/\beta} \quad (t\in B). \tag{5.25}$$

Proof. It follows from (c) and Theorem 4.5 that the function

$$f(t) = \exp(-\rho(t)^\alpha) \qquad (t\in B)$$

is positive definite on B, and hence, by (a), that $f(\lambda_1 t_1 + \lambda_2 t_2 + \cdots + \lambda_n t_n)\in\mathrm{PD}(\mathbb{R}^n)$ for $n\geq 1$ and every choice of points $t_1,t_2,\dots,t_n$ in B. Since $f(0) = 1$ this implies that $f(\lambda_1 t_1 + \lambda_2 t_2 + \cdots + \lambda_n t_n)$ is the Fourier transform of a probability measure

$P_{t_1, t_2 \ldots t_n}$ on R^n (a positive Borel measure of total mass 1). It is clear that this collection of probability measures satisfy the following compatibility conditions:

C_1. $P_{t_1, t_2 \ldots, t_n}(A) = P_{\pi(t_1, t_2, \ldots, t_n)}(\pi A)$ where π is any permutation of $t_1, t_2, \ldots, t_n$ and πA is the set of permutations of the coordinates of the Borel set A under π.

C_2. $P_{t_1, \ldots, t_n, t_{n+1}, \ldots, t_{n+m}}(A^{(n)} \times \mathbb{R}^m) = P_{t_1, \ldots, t_n}(A^{(n)})$
for any Borel set $A^{(n)} \subset \mathbb{R}^n$.

Thus we may invoke a theorem of Kolmogorov [30] which shows how to construct a probability space (Ω, P) and random variables $Y_t (t \in B)$ such that

$$f(\lambda_1 t_1 + \lambda_2 t_2 + \cdots + \lambda_n t_n) = \int_{\Omega} \exp(i \sum_{j=1}^{n} \lambda_j Y_{t_j}) dP \qquad (5.26)$$

$$= E(\exp(i \sum_{j=1}^{n} \lambda_j Y_{t_j}))$$

for all $n \geq 1$, $(\lambda_1, \lambda_2, \ldots, \lambda_n) \in \mathbb{R}^n$ and points $t_1, t_2, \ldots, t_n \in B$. (Here "random variable" and "E = expection" have their usual meanings.)

In order to show that the map $t \to Y_t$ is linear, choose $a, \lambda \in \mathbb{R}$ and $t \in B$ and note that

$$E(\exp i\lambda(Y_{at} - aY_t)) = f(\lambda(at) - (\lambda a)t) = 1.$$

Hence

$$\lambda(Y_{at} - aY_t) = 0 \pmod{2\pi} \text{ a.e.,}$$

and since this holds for all $\lambda \in \mathbb{R}$, we conclude that

$$Y_{at} = aY_t \text{ a.e.} \qquad (a \in \mathbb{R}, t \in B).$$

In the same way the relation

$$E(\exp(i\lambda(Y_{s+t} - Y_s - Y_t))) = f(\lambda(s+t) - \lambda s - \lambda t) = 1$$

implies that

$$Y_{s+t} = Y_s + Y_t \text{ a.e.} \qquad (s \in B, t \in B).$$

For every $\lambda \in \mathbb{R}$ and $t \in B$, we have

$$E(\exp(i\lambda Y_t)) = f(\lambda t) = \exp(-|\lambda|^\alpha \rho(t)^\alpha) \qquad (5.27)$$

by virtue of (5.26) and (b). We know already (see Corollary 4.9) that $\exp(-|\lambda|^\alpha)$ $(0 < \alpha \leq 2)$ is the Fourier transform of a probability measure. But much more is true: there exists a nonnegative summable function φ_α on $\mathbb{R}$ such that

$$\exp(-|\lambda|^\alpha) = \int_{-\infty}^{\infty} e^{-i\lambda u} \varphi_\alpha(u) du \qquad (\lambda \in \mathbb{R}) \qquad (5.28)$$

and

$$C_{\alpha\beta} = \int_{-\infty}^{\infty} |u|^\beta \varphi_\alpha(u) du < \infty \text{ for } 0 < \beta < \alpha < 2$$

and for all β if $\alpha = 2$.

Only that part of the conclusion relating to $\alpha = 2$ is clear and it follows from the formula (4.7). A proof for $\alpha < 2$ will be found in [47, p.179].

Combining (5.27) and (5.28) we obtain

$$E(\exp(i\lambda Y_t)) = \int_{-\infty}^{\infty} e^{-i\lambda u} \, \phi_\alpha(u/\rho(t)) \frac{du}{\rho(t)} \qquad (\lambda \in \mathbb{R}),$$

from which we conclude that $\varphi_\alpha(u/\rho(t)) \, du/\rho(t)$ is the probability law of Y_t. Therefore

$$E(|Y_t|^\beta) = \int_{-\infty}^{\infty} |u|^\beta \, \varphi_\alpha(u/\rho(t)) \, du/\rho(t) = C_{\alpha\beta} \, \rho(t)^\beta.$$

It is now clear that if we put $X_t = C_{\alpha\beta}^{-1/\beta} \, Y_t$, the map $\phi : t \to X_t$ $(t \in B)$ is the required linear isometry. $\square$

There is an immediate corollary.

Corollary 5.8. *If $0 < p \leq 2$ and $\|x - y\|^{p/2}$ is of negative type on B, then B is linearly isometric to a subspace of an L^q space, $0 < q < p$. Further, if $p = 2$ then B is linearly isometric to a subspace of some L^q space, for all $0 < q < \infty$.*

This last statement, concerning $p = 2$, generalizes an old theorem of Banach and Mazur [4, p.203] to the effect that L^2 is linearly isomorphic with a closed subspace of L^q for $q > 1$.

Lemma 5.9. *Suppose that $0 < \alpha \leq 2$ and ρ is a continuous function from $\mathbb{R}^n$ $(n \geq 1)$ into $\mathbb{R}^+$ which satisfies conditions (b) and (c) of the previous lemma. Then there exists a positive measure μ on the unit sphere S_{n-1} in $\mathbb{R}^n$ such that*

$$\rho(t) = \left(\int_{S_{n-1}} |(t,u)|^\alpha \, d\mu(u) \right)^{1/\alpha} \qquad (t \in \mathbb{R}^n). \tag{5.29}$$

Proof. By the preceding lemma there exists a probability space (Ω, P) and, corresponding to each β, $0 < \beta < \alpha$, a linear map $\phi : t \to X_t$ from $\mathbb{R}^n$ into $L^\beta(\Omega, P)$ such that

$$E(|X_t|^\beta)^{1/\beta} = \rho(t) \qquad (t \in \mathbb{R}^n).$$

Let $e_1, e_2, \ldots, e_n$ denote the standard basis in $\mathbb{R}^n$ and define a measure Q on Ω by

$$dQ(u) = [X_{e_1}^2 \, (u) + X_{e_2}^2 \, (u) + \cdots + X_{e_n}^2 \, (u)]^{\beta/2} \, dP(u).$$

Since $\beta/2 < 1$,

$$\int_\Omega dQ(u) \leq \int_\Omega [|X_{e_1} \, (u)|^\beta + |X_{e_2} \, (u)|^\beta + \cdots + |X_{e_n} \, (u)|^\beta] dP(u)$$

$$= \rho(e_1)^\beta + \rho(e_2)^\beta + \cdots + \rho(e_n)^\beta, \tag{5.30}$$

so Q is a finite measure.

The linearity of ϕ implies that

$X_x = x_1 X_{e_1} + x_2 X_{e_2} + \cdots + x_n X_{e_n}$ when $x = x_1 e_1 + x_2 e_2 + \cdots + x_n e_n$ $(x_1, x_2, \ldots, x_n \in \mathbb{R})$, and therefore $X_{e_1}^2 \, (u) + X_{e_2}^2 \, (u) + \cdots + X_{e_n}^2 \, (u) = 0$ implies that $X_x(u) = 0$. Hence we can define

$$Y_x(u) = [X_{e_1}^2 \, (u) + X_{e_2}^2 \, (u) + \cdots + X_{e_n}^2 \, (u)]^{-1/2} \, X_x(u) \qquad (x \in \mathbb{R}^n).$$

And clearly,

$$\int_\Omega |Y_x(u)|^\beta \, dQ(u) = \int_\Omega |X_x(u)|^\beta \, dP(u) = \rho(x)^\beta.$$

In terms of the map $\tau : u \to (Y_{e_1}(u), Y_{e_2}(u), \dots, Y_{e_n}(u))$ from Ω to $\mathbb{R}^n$, we define the positive measure μ_β on $\mathbb{R}^n$ by

$$\mu_\beta(A) = Q(\tau^{-1}(A))$$

for each Borel set $A \subset \mathbb{R}^n$. Since $Y_{e_1}^2 + Y_{e_2}^2 + \cdots + Y_{e_n}^2 = 1$ almost everywhere relative to Q, the measure μ_β is concentrated on S_{n-1}. According to (5.30) the measures μ_β are bounded for $0 < \beta < \alpha$, so as $\beta \mathord{\uparrow} \alpha$ we may conclude that the μ_β converge to a limit μ in the weak* topology of Borel measures on S_{n-1} considered as the dual space of $C(S_{n-1})$.

For every $t = \lambda_1 e_1 + \lambda_2 e_2 + \cdots + \lambda_n e_n$ $(\lambda_1, \lambda_2, \dots, \lambda_n) \in \mathbb{R}^n$,

$$\int_{S_{n-1}} |(t,x)|^\beta d\mu_\beta(x) = \int_{S_{n-1}} |\lambda_1 x_1 + \lambda_2 x_2 + \cdots + \lambda_n x_n|^\beta \, d\mu_\beta(x)$$

$$= \int_\Omega |\lambda_1 Y_{e_1} + \lambda_2 Y_{e_2} + \cdots + \lambda_n Y_{e_n}|^\beta dQ$$

$$= \int_\Omega |Y_t|^\beta dQ = \rho(t)^\beta.$$

Letting $\beta \mathord{\uparrow} \alpha$ we get (5.29) as required. $\square$

Theorem 5.10. *If $\|x - y\|^{p/2}, 0 < p \le 2$, is of negative type on the finite-dimensional normed linear space B, then there is a linear isometry from B into some L^p space.*

Proof. Identify B with $\mathbb{R}^n$ and apply the preceding lemma with $\rho(t) = \|t\|$ and ϕ the map that sends $t \in \mathbb{R}^n$ to the function $\phi(t)(x) = (t,x)$ $(x \in S_{n-1})$. $\square$

Proof of Theorem 5.1. The proof is now evident. For if B is a normed linear space such that $\|x - y\|^{p/2}$ is of negative type for some p satisfying $1 \le p \le 2$, then each of its finite-dimensional subspaces is of type p by Theorem 5.10, and hence so is B by Theorem 5.5. $\square$

We close this section with an application of Corollary 5.6 on metric embedding in L^q spaces which extends (4.14).

Theorem 5.11. *If $1 \le p \le q \le 2$, then $L^p(\Omega, \mu) = L^p$ with the norm $\|x - y\|_p^\alpha$ is of type q if and only if $0 < \alpha \le p/q$.*

Proof. Suppose that $\alpha > 0$ and that the metric space $(L^p, \|x - y\|_p^\alpha)$ is of type q. Since $\|x - y\|_q^{q/2}$ is of negative type on L^q (see (4.13)), L^q with the metric $\|x - y\|_q^{q/2}$ is of type 2, that is, embeddable in L^2. Hence L^p with the metric $\|x - y\|_p^{(\alpha q)/2}$ is embeddable in L^2. In view of Corollary 4.11 this is impossible unless $(\alpha q)/2 \le p/2$ or $\alpha \le p/q$. Thus the restriction on α is necessary.

Turning to the sufficiency, let $\mathscr{E}$ denote the dense subspace of L^p consisting of the simple measurable functions. It clearly suffices to show that the linear metric space $(\mathscr{E}, \|x - y\|_p^\alpha)$ can be embedded in some L^q space and such will be the case, according to Corollary 5.6, if every finite-dimensional subspace can be so embedded. As every finite-dimensional subspace of $\mathscr{E}$ is isometric to some

ℓ_n^p, we need only show that the linear metric space $(\ell_n^p, \|x-y\|_p)$ is of type q.

First, suppose that $\alpha = p/q$. The real line with the metric $|x-y|^{p/q}$ can be embedded in L^2 and L^2 is of type q by Lemma 5.7. Hence there is a map ϕ_1 from $\mathbb{R}$ into some $L^q(N_1, v_1)$ such that

$$|x_1 - y_1|^{p/q} = \|\phi_1(x_1) - \phi_1(x_2)\|_q \qquad (x_1, y_1 \in \mathbb{R}).$$

Let (N, v) denote the measure space formed by the direct sum of the measure spaces $(N_1, v_1), \ldots, (N_n, v_n)$, each (N_j, v_j) being a copy of $(N_1, v_1)(j=2,\ldots,n)$, and let ϕ_j have the action of ϕ_1 $(j=2,3,\ldots,n)$. The map J from $\mathbb{R}^n$ to $L^q(N, v)$ defined by

$$J : (x_1, x_2, \ldots, x_n) \to \sum_{j=1}^{n} \phi_j(x_j)$$

satisfies

$$|x_1 - y_1|^p + |x_2 - y_2|^p + \cdots + |x_n - y_n|^p = \sum_{j=1}^{n} \|\phi_j(x_j) - \phi_j(y_j)\|_q^q$$

$$= \|J(x_1, x_2, \ldots, x_n) - J(y_1, y_2, \ldots, y_n)\|_q^q.$$

Hence $(\ell_n^p, \|x-y\|_p^{p/q})$ is of type q.

When $0 < \alpha < p/q$ or $0 < q < p/\alpha$, consider the cases $p/\alpha > 2$ and $p/\alpha \leq 2$. In the first case $\alpha < p/2$, so $(L^p, \|x-y\|_p^\alpha)$ is embeddable in L^2 by (4.14) and L^2 is of type q by Lemma 5.7. In the second case, apply the initial reasoning to see that $(L^p, \|x-y\|_p^\alpha)$ is of type p/α. Since $q < p/\alpha \leq 2$, $L^{p/\alpha}$ is embeddable in L^q by Corollary 5.8. $\square$

Chapter II. The Classes N(X) and RPD(X): Integral Representations

This chapter continues the study of radial positive definite functions and functions of negative type begun in the preceding chapter. Following methods introduced by Schoenberg [79,80,81] and recently refined by Bretagnolle, Dacunha Castelle, and Krivine [11], and Kuelbs [52], we are going to present characterizations of the classes RPD(X) and N(X) when X is one of the spaces $\mathbb{R}^n$ or L^p ($n=1,2,\ldots$; $0 < p \le \infty$).

These characterizations are in terms of integral transforms directly related to either a Fourier or Laplace transform. The functions in RPD($\mathbb{R}^n$) are just the radial Fourier transforms, but the representations for the class RPD(L^p), when the underlying space is infinite-dimensional, depend on a subtle relationship between positive definite and completely monotonic functions. In general the characterization of N(X) follows from that of RPD(X) through a limiting procedure; thus the concept of positive definiteness is dominant. However, we do present a characterization of N($\mathbb{R}^n$) due to von Neumann which uses operator methods to exploit the associated isometric embedding in Hilbert space [66].

§6. Radial Positive Definite Functions on $\mathbb{R}^n$

Fix a positive integer n. Each element F of RPD($\mathbb{R}^n$) defines a function f on $\mathbb{R}^n$ by

$$f(x) = F(\|x\|) \qquad (x \in \mathbb{R}^n) \tag{6.1}$$

which is positive definite in the sense of (4.1) and radial, that is, $f \circ J = f$ for any isometry J of $\mathbb{R}^n$ leaving the origin fixed. Consequently, any positive definite function on $\mathbb{R}^n$ which is radial defines an element F of RPD($\mathbb{R}^n$) via (6.1). In this way the class RPD($\mathbb{R}^n$) may be identified with those radial functions on $\mathbb{R}^n$ which are Fourier transforms of finite positive measures. We intend to exploit this symmetry through a special kernel function.

Let S_{n-1} be the unit sphere in $\mathbb{R}^n$:

$$S_{n-1} = \{x \in \mathbb{R}^n : x_1^2 + x_2^2 + \cdots + x_n^2 = 1\}.$$

On this compact set σ_{n-1} will denote surface measure. In terms of spherical coordinates for R^n,

$$x_1 = r \cos(\theta_1) \qquad\qquad\qquad 0 \le r < \infty,$$
$$x_2 = r \sin(\theta_1) \cos(\theta_2) \qquad\qquad 0 \le \theta_1 \le \pi,$$

$$x_3 = r \sin(\theta_1) \sin(\theta_2) \cos(\theta_3) \qquad\qquad 0 \le \theta_2 \le \pi,$$

.

.

.

$$x_{n-1} = r \sin(\theta_1) \sin(\theta_2) \ldots \sin(\theta_{n-2}) \cos(\theta_{n-1}) \qquad 0 \le \theta_{n-2} \le \pi,$$
$$x_n = r \sin(\theta_1) \sin(\theta_2) \ldots \sin(\theta_{n-2}) \sin(\theta_{n-1}) \qquad 0 \le \theta_{n-1} \le 2\pi,$$

and

$$d\sigma_{n-1} = \sin^{n-2}\theta_1 \ldots \sin^2\theta_{n-3} \sin\theta_{n-2}\, d\theta_1 \ldots d\theta_{n-1}. \tag{6.2}$$

The σ_{n-1} measure of S_{n-1} is given by

$$\omega_{n-1} = \int_{S_{n-1}} d\sigma_{n-1} = \int_0^{2\pi} \ldots \int_0^{\pi} \sin^{n-2}\theta_1 \ldots \sin\theta_{n-2} d\theta_1 \ldots d\theta_{n-1} \tag{6.3}$$

$$= 2\pi \prod_{j=2}^{n-1} \int_0^{\pi} \sin^{n-j}\theta_{j-1} d\theta_{j-1} = \frac{2\pi^{n/2}}{\Gamma(n/2)}.$$

These computations are straightforward.

Definition 6.1. *For each positive integer n, Ω_n is the function on $\mathbb{R}^n$ defined by*

$$\Omega_n(x) = \omega_{n-1}^{-1} \int_{S_{n-1}} e^{i(x,\xi)} d\sigma_{n-1}(\xi) \quad (x \in \mathbb{R}^n). \tag{6.4}$$

Thus $\Omega_n(x)$ is the mean-value of $e^{i(x,\xi)}$ over S_{n-1}. When $n = 1$, $S_0 = \{-1,1\}$, σ_0 is the measure consisting of a point mass of $1/2$ at each of -1 and 1 and

$$\Omega_1(x) = \cos(x) \qquad (x \in \mathbb{R}).$$

When $n = 2$,

$$\Omega_2(x) = \frac{1}{2} \int_0^{2\pi} e^{i(x_1 \cos\theta + x_2 \sin\theta)}\, d\theta = J_0(|x|) \qquad (x \in \mathbb{R}^2).$$

In general Ω_n is a radial function,

$$\Omega_n(x) = \Omega_n(\|x\|) \qquad (x \in \mathbb{R}^n). \tag{6.5}$$

This is a consequence of the fact that σ_{n-1} is invariant under isometries of $\mathbb{R}^n$ which leave the origin fixed. With this observation we put

$$\Omega_n(r) = \Omega_n((r,0,0,\ldots,0)) \qquad (r \ge 0). \tag{6.6}$$

Combining (6.6) and (6.3), we obtain, for $n \ge 2$, the representation

$$\Omega_n(r) = \omega_{n-1}^{-1} \int_{S_{n-1}} e^{ir\xi_1} d\sigma_{n-1}(\xi)$$

which, upon changing to polar coordinates, reduces to

$$\Omega_n(r) = \omega_{n-1}^{-1} \omega_{n-2} \int_0^{\pi} e^{ir\cos\theta} \sin^{n-2}\theta\, d\theta \tag{6.7}$$

$$= \frac{\int_0^{\pi} e^{ir\cos\theta} \sin^{n-2}\theta\, d\theta}{\int_0^{\pi} \sin^{n-2}\theta\, d\theta}.$$

From this formula we easily deduce that

$$\Omega_3(r) = (\sin r)/r \qquad (r \geq 0). \tag{6.8}$$

It is clear from (6.7) that Ω_n is the restriction to $\mathbb{R}^+$ of an entire function of r and thus it has a power series expansion valid for all r which can be calculated by expanding the exponent in the integrand of (6.7) and integrating term by term:

$$\Omega_n(r) = \sum_{p=0}^{\infty} (-1)^p \frac{1}{2^p p! n(n+2)\cdots(n+2p-2)} r^{2p}. \tag{6.9}$$

A comparison of this expansion with the series for the Bessel function of order n leads to the formula

$$\Omega_n(r) = \Gamma(n/2) (2/r)^{(n-2)/2} J_{(n-2)/2}(r). \tag{6.10}$$

Theorem 6.2. *A function F is radial positive definite on $\mathbb{R}^n$, $1 \leq n < \infty$, if and only if there exists a finite positive measure α on $\mathbb{R}^+$ such that*

$$F(r) = \int_0^\infty \Omega_n(ru) \, d\alpha(u) \qquad (0 \leq r < \infty). \tag{6.11}$$

Proof. A function F given by (6.11), with a positive and finite α, is continuous and bounded since Ω_n shares the same properties. For each fixed u, $\Omega_n(ru)$ is radial positive definite on $\mathbb{R}^n$ by (6.4) and (6.5); hence F is a pointwise limit of functions in RPD($\mathbb{R}^n$) and, consequently, is itself a member of RPD($\mathbb{R}^n$) by Theorem 4.4(d).

For the converse, suppose that $F \in$ RPD($\mathbb{R}^n$). As we have already observed the function $f(x) = F(\|x\|)$ is positive definite on $\mathbb{R}^n$, and so has a representation

$$f(x) = \int_{\mathbb{R}^n} e^{i(x,u)} \, dv(u) \qquad (x \in \mathbb{R}^n) \tag{6.12}$$

for some finite positive measure v on $\mathbb{R}^n$. Fix $r \geq 0$. Since $f(r\xi)$ is constant as ξ ranges over S_{n-1}, we may write

$$F(r) = f((r,0,\ldots,0)) = \omega_{n-1}^{-1} \int_{S_{n-1}} f(r\xi) \, d\sigma_{n-1}(\xi).$$

For the integrand substitute from (6.12) and use the representation (6.4) to obtain

$$F(r) = \omega_{n-1}^{-1} \int_{S_{n-1}} \left\{ \int_{\mathbb{R}^n} e^{i(r\xi,u)} \, dv(u) \right\} d\sigma_{n-1}(\xi)$$

$$= \int_{\mathbb{R}^n} \left\{ \omega_{n-1}^{-1} \int_{S_{n-1}} e^{i(\xi,ru)} \, d\sigma_{n-1}(\xi) \right\} dv(u)$$

$$= \int_{\mathbb{R}^n} \Omega_n(ru) \, dv(u) = \int_0^\infty \Omega_n(ru) \, d\alpha(u) \qquad (r \geq 0).$$

The reduction in the last step from an integral over $\mathbb{R}^n$ to an integral from 0 to ∞ is possible because the integrand is radial. The definition of α on Borel subsets G of $[0, \infty)$ is

$$\alpha(G) = v(\{x \in \mathbb{R}^n : \|x\| \in G\}). \quad \square$$

§7. Positive Definite Functions on Infinite-Dimensional Linear Spaces

Since $\mathbb{R} \subset \mathbb{R}^2 \subset \cdots \subset H$ we have the inclusion RPD $(\mathbb{R}) \supset$ RPD($\mathbb{R}^2$) $\supset \cdots \supset$ RPD(H) from which it follows that

$$\text{RPD}(H) = \bigcap_{n \geq 1} \text{RPD}(\mathbb{R}^n),$$

in the sense of set intersection. This symbolizes the obvious: a function is radial positive definite on H if and only if it is radial positive definite on every finite-dimensional subspace of H. But it does suggest that a characterization of RPD(H) might emerge by letting n tend to infinity in the integral formula (6.11), and such is indeed the case [81]. However, we prefer to present the direct attack of Kuelbs [52] which exposes a fundamental relationship between radial positive definite functions on H and completely monotonic functions. We begin with some definitions and assumptions.

Throughout this section we shall assume that E is an infinite-dimensional linear space, real or complex, and that θ is a map from E to $\mathbb{R}^+$ for which the following conditions are satisfied:

$$\theta(-x) = \theta(x), \ \theta(0) = 0, \ \theta(x) \geq 0 \qquad (x \in E). \tag{7.1}$$

There exists in E a linearly independent set $e_1, e_2, \dots$ such that θ is additive, (7.2) $\theta(x+y) = \theta(x) + \theta(y)$, on the linear span $S = \text{sp}\{e_1, e_2, \dots\}$ with respect to those pairs of elements x and y in S generated by disjoint collections of the e_j's.

For each $t > 0$ and positive integer j there exist $t_j > 0$ such that $\theta(t_j e_j) = t$. (7.3)

The conditions imposed can be realized in many classical Banach spaces by taking θ to be an appropriate power of the norm. For example, in Hilbert space one can take θ to be the square of the norm and the e_j's to be an orthonormal basis; in the space $L^p(\Omega, \mu)$ $(0 < p < \infty)$ corresponding to a measure space (Ω, μ) which contains infinitely many pairwise disjoint sets $\{E_j\}$ of positive but finite μ-measure, one can set

$$\theta(f) = \int_\Omega |f|^p d\mu \qquad (f \in L^p)$$

and take e_j = characteristic function of E_j, $j = 1, 2, \dots$.

Recall that a function f is *completely monotonic* on $(0, \infty)$ provided that

$$(-1)^k f^{(k)}(x) \geq 0 \ \text{ for } \ k = 0, 1, 2, \dots \ \text{ and } \ 0 < x < \infty; \tag{7.4}$$

or, alternatively, that

$$\Delta_h^n f(x) = \sum_{k=0}^n (-1)^k \binom{n}{k} f(x+kh) \geq 0 \ \text{ for } \ n = 0, 1, 2, \dots \tag{7.5}$$

and for all nonnegative numbers x and h. Further, f is *completely monotonic on* $[0, \infty)$ provided it is continuous there and satisfies (7.4) or (7.5).

For a proof of the equivalence of (7.4) and (7.5), of the following fundamental result, and a discussion of related topics, see D. V. Widder [89].

Theorem 7.1. *The completely monotonic functions on* $(0, \infty)$ *are precisely those which admit a representation*

$$f(x) = \int_0^\infty e^{-xu} d\alpha(u) \qquad (0 < x < \infty) \tag{7.6}$$

for some positive (finite on compact subsets) measure α *on* $\mathbb{R}^+$. *The representing* *measure is unique. Further, f is completely monotonic on $\mathbb{R}^+$ if and only if (7.6) holds with α a finite and positive measure on $\mathbb{R}^+$.*

Theorem 7.2.   *If the function f on $\mathbb{R}^+$ is such that $f \circ \theta$ is positive definite on E, then f is completely monotonic on $(0,\infty)$.*

Proof.   The first step is to show that $f(t) \geq 0$ for $t > 0$. To do this choose t_j in (7.3) so that $\theta(t_j e_j) = t/2$ and put $x_j = t_j e_j$ for $j = 1,2,\ldots$. Since $f \circ \theta$ is positive definite on E, we have

$$0 \leq \sum_{j,k=1}^{N} f(\theta(x_j - x_k)) = Nf(0) + N(N-1)\,f(t).$$

The required inequality is apparent upon first dividing both members by N^2 and then letting $N \to \infty$.

Next we show that the function $\Delta_h^1 f\,(h > 0)$ defined by $\Delta_h^1 f(x) = f(x) - f(x+h)$, $0 \leq x < \infty$, is such that $\Delta_h^1 f \circ \theta$ is positive definite on S and consequently, by repetition of the argument in the first paragraph, that $\Delta_h^1 f(t) \geq 0$ for $t > 0$. It will then follow by induction that the function

$$\Delta_h^n f \circ \theta = \Delta_h^1 (\Delta_h^{n-1} f) \circ \theta$$

is positive definite on S for $n = 2,3,\ldots$ and that (7.5) holds.

To carry out this second step fix a positive integer n, choose points $x_1, x_2, \ldots, x_n$ in S and real numbers $\xi_1, \xi_2, \ldots, \xi_n$. Choose an integer ℓ so large that each $x_j, 1 \leq j \leq n$, is contained in the linear span of $e_1, e_2, \ldots, e_\ell$, then take $m > \ell$ and $s > 0$ so that $\theta(s\,e_m) = h$, h being positive and fixed. Define

$$y_j = \begin{cases} x_j, & 1 \leq j \leq n \\ x_{j-n} + s e_m, & n < j \leq 2n \end{cases} \quad \text{and } \eta_j = \begin{cases} \xi_j, & 1 \leq j \leq n \\ -\xi_{j-n}, & n < j \leq 2n \end{cases}$$

The inequality $0 \leq \sum_{j,k=1}^{2n} f(\theta(y_j - y_k))\eta_j \eta_k$ splits naturally into

$$0 \leq \sum_{j,k=1}^{n} f(\theta(x_j - x_k))\xi_j \xi_k + \sum_{j,k=n+1}^{2n} f(\theta(x_{j-n} - x_{k-n}))\xi_{j-n}\xi_{k-n}$$

$$- \sum_{j=1,k=n+1}^{n,2n} f(\theta(x_j - x_{k-n}) + h)\xi_j \xi_{k-n} - \sum_{j=n+1,k=1}^{2n,n} f(\theta(x_{j-n} - x_k) + h)\xi_{j-n}\xi_k,$$

and these sums combine by pairs to give

$$0 \leq 2\sum_{j,k=1}^{n} [f(\theta(x_j - x_k)) - f(\theta(x_j - x_k) + h)]\xi_j \xi_k = 2\sum_{j,k=1}^{m} \Delta_h^1 f \circ \theta(x_j - x_k)\xi_j \xi_k.$$

Hence $\Delta_h^1 f \circ \theta$ is positive-definite on S. $\square$

Definition 7.3.   *Let f and g be functions on E. The statement that f is g-symmetric means that $g(x) = g(y)$ implies that $f(x) = f(y)$.*

Theorem 7.4.   *Suppose Φ is positive definite and θ-symmetric on E. Then there exists a finite and positive measure α on $\mathbb{R}^+$ such that*

$$\Phi(x) = \int_0^\infty e^{-u\theta(x)}\,d\alpha(u) \qquad (x \in E,\ \theta(x) \neq 0). \tag{7.7}$$

Proof. It follows from (7.1) and (7.3) that every nonnegative number is in the range of θ. This observation and the θ-symmetry of Φ imply that the association $\theta(x) \to \Phi(x)$ $(x \in E)$ defines a function f on $\mathbb{R}^+$ such that $f(\theta(x)) = \Phi(x)$. Since Φ is positive definite on E we may call upon Theorems 7.1 and 7.2 to establish the existence of a positive measure α on $\mathbb{R}^+$ such that

$$f(t) = \int_0^\infty e^{-tu} \, d\alpha(u) \qquad (t > 0).$$

The obvious substitution gives (7.7) and the finiteness of α follows from the inequality

$$f(0) \geq \lim_{t \downarrow 0} f(t) = \lim_{t \downarrow 0} \int_0^\infty e^{-tu} \, d\alpha(u) \geq \int_0^\infty d\alpha(u). \qquad \square$$

Note. The reason we cannot claim equality in (7.7) on the set $\theta^{-1}(0) = \{x \in E : \theta(x) = 0\}$ is that there is no continuity assumption ab initio on Φ. Hence Φ will continue to be positive definite on E even when its constant value on $\theta^{-1}(0)$ is increased.

These results provide a characterization of certain of the spaces $\mathrm{RPD}(L^p)$. Let (Ω, μ) be a measure space and put $L^p(\mu) = L^p(\Omega, \mu)$ $(0 < p \leq \infty)$.

Theorem 7.5. *Suppose that $L^p(\mu)$ $(0 < p \leq 2)$ has infinite (linear) dimension. Then F is radial positive definite on $L^p(\mu)$ $(0 < p \leq 2)$ if and only if there exists a finite positive measure α on $\mathbb{R}^+$ such that*

$$F(r) = \int_0^\infty e^{-r^r u} \, d\alpha(u) \qquad (r \geq 0). \tag{7.8}$$

Proof. First we establish the existence in Ω of an infinite sequence $\{E_j\}$ of pairwise disjoint sets of finite and positive μ-measure. To do this suppose that such sets $E_1, E_2, \ldots, E_k$ have been constructed. The infinite-dimensionality of $L^p(\mu)$ guarantees the existence of a function y in $L^p(\mu)$ which either (a) is not constant μ-almost everywhere on some E_j $(1 \leq j \leq k)$, or (b)

$$\int_F |y|^p \, d\mu > 0,$$

where $F = \Omega \setminus \bigcup_{j=1}^k E_j$. If (a) holds, one of the sets E_j splits into two sets of positive measure and, if (b) holds, F contains a set of positive and finite μ measure. The existence of the sequence now follows by induction. Put $\theta(x) = \|x\|_p^p$ $(x \in L^p(\mu))$ and let e_j be the characteristic function of E_j, $j = 1, 2, \ldots$.

According to (4.12) the function $e^{-\|x - y\|_p^p u}$ $(u > 0)$ is positive definite on L^p for $0 < p \leq 2$. Hence every F defined by the integral formula (7.8), relative to a positive and finite measure, must be radial positive definite on $L^p(\mu)$.

For the converse, suppose $F \in \mathrm{RPD}(L^p)$. Put $\Phi(x) = F(\|x\|_p)$ and note that Φ is θ-symmetric. In view of our opening remarks Theorem 7.4 applies and (7.8) follows from (7.7). $\square$

There is an important corollary, due to Schoenberg in the case $p = 2$ [81].

Corollary 7.6. *Fix $p \in (0, 2]$ and suppose $L^p(\mu)$ is infinite-dimensional. A function f is completely monotonic on $\mathbb{R}^+$ if and only if the function $F(r) = f(r^p)$ is radial positive definite on $L^p(\mu)$.*

We presented Theorem 7.5 at this juncture because it illuminates at once the fundamental nature of Theorem 7.4 and, besides, it was one of the stated goals of the chapter. But it is silent on the situation for $2 < p \leq \infty$. Continuing with the results of Kuelbs, we consider this extended range of p in a generalized setting.

Let ω be a fixed nonnegative, nondecreasing, and continuous function on $\mathbb{R}^+$ such that

$$\omega(0) = 0, \ \omega(t) > 0 \ \text{for} \ t > 0, \ \lim_{t \to \infty} \omega(t) = \infty, \qquad (7.9)$$

and

$$\omega(2t) \leq K \, \omega(t) \ \text{for} \ t \geq 0 \ \text{and some fixed} \ K. \qquad (7.10)$$

Given a measure space (Ω, μ), define $L^\omega = L^\omega(\Omega, \mu)$ to be the space of all complex μ-measurable functions on Ω such that

$$\rho(f) = \int_\Omega \omega(|f(u)|) \, d\mu(u) < \infty.$$

Condition (7.10) implies that $\omega(|s+t|) \leq K[\omega(|s|) + \omega(|t|)]$ and hence that L^ω is a linear space. In particular, the L^p spaces arise by taking $\omega(t) = |t|^p$, $0 < p < \infty$.

Theorem 7.7. *Suppose that L^ω is infinite-dimensional. Then every ρ-symmetric and positive definite function Φ on L^ω has the form*

$$\Phi(f) = \int_0^\infty e^{-u\rho(f)} \, d\alpha(u) \qquad (\rho(f) > 0) \qquad (7.11)$$

for some finite and positive measure α on $\mathbb{R}^+$.

Proof. The condition (7.9) on ω together with the assumption on the linear dimension of L^ω guarantees (as in the proof of Theorem 7.5) the existence of an infinite sequence $\{E_j\}$ of pairwise disjoint sets in Ω of finite and positive μ-measure. Put $\theta(f) = \rho(f)$ and let e_j be the characteristic function on E_j. Again Theorem 7.4 applies so that (7.11) follows from (7.7). $\square$

There is no assertion made here that a function Φ given by (7.11) can ever be positive definite on L^ω in other than the trivial situation which occurs when α is a point mass at zero. As we shall now show the situation is, essentially, that this trivial case holds or else $\omega \in N(\mathbb{R})$.

Theorem 7.8. *Suppose that ω is as in Theorem 7.7 and that Ω contains infinitely many disjoint sets of equal finite and nonzero measure. Let Φ be given by (7.11) with respect to a measure α which is not concentrated at 0 and either*
  (a) $\int_0^\infty e^{-u\lambda\rho(f)} \, d\alpha(u)$ *is positive definite on L^ω for a sequence of positive λ's tending to 0, or*
  (b) $\alpha((0,\varepsilon)) > 0$ *for each $\varepsilon > 0$.*
Then the following are equivalent:
  (1) $\Phi(f)$ *is positive definite on L^ω.*
  (2) $e^{-\lambda\rho(f)}$ *is positive definite on L^ω for each $\lambda > 0$.*
  (3) $e^{-\lambda\omega(|t|)}$ *is positive definite on $\mathbb{R}$ for each $\lambda > 0$.*
  (4) $\omega(|s-t|)^{1/2}$ *is a function of negative type on $\mathbb{R}$.*

Proof. The equivalence of (3) and (4) is the content of Theorem 4.6. To see that (4) implies (2), let $f_1, f_2, \ldots, f_n$ be elements in L^ω and choose real c_j such that $\Sigma c_j = 0$. Then

$$\sum_{j,k=1}^{n} c_j c_k \, \rho(f_j - f_k) = \int_{\Omega} \sum_{j,k=1}^{n} c_j c_k \omega(|f_j(u) - f_k(u)|) d\alpha(u) \le 0,$$

since $\omega^{1/2}$ is of negative type on R. Hence $\rho^{1/2}$ is a quasi-metric of negative type on L^ω and (2) follows from Theorem 4.5. Condition (1) is an easy consequence of (2). The remainder of the proof is devoted to showing that (1) implies (4).

We lose nothing by assuming that $\int_0^\infty d\alpha = 1$. Choose in Ω an infinite sequence of pairwise disjoint sets $A_1, A_2, \ldots$ such that

$$0 < \mu(A_j) = \delta < \infty \quad (j = 1, 2, \ldots),$$

and let $f_1, f_2, \ldots$ denote their characteristic functions. For each positive integer n and numbers $t_1, t_2, \ldots, t_n \in \mathbb{R}$ define

$$\begin{aligned}
\Phi_n(t_1, t_2, \ldots, t_n) &= \int_0^\infty \exp\left(-u\rho\left(\sum_1^n t_j f_j\right)\right) d\alpha(u) \\
&= \int_0^\infty \exp\left(-u\delta \sum_1^n \omega(|t_j|)\right) d\alpha(u).
\end{aligned} \tag{7.12}$$

Then, by (1), Φ_n is positive definite on $\mathbb{R}^n$ and hence it is the Fourier transform of a probability measure P_n on $\mathbb{R}^n$. Since $\Phi_{n+1}(t_1, t_2, \ldots, t_n, 0) = \Phi_n(t_1, t_2, \ldots, t_n)$ the measures P_n form a compatible collection (see Lemma 5.7), so it follows from the theorem of Kolmogorov that there exists a probability space $(\psi, \mathscr{A}, P)$ and a sequence of random variables $X_1, X_2, \ldots$ on ψ such that

$$\Phi_n(t_1, t_2, \ldots, t_n) = E(e^{i(t_1 X_1 + t_2 X_2 + \cdots + t_n X_n)}) \tag{7.13}$$

for every choice of $(t_1, t_2, \ldots, t_n) \in \mathbb{R}^n$ $(n \ge 1)$. From (7.12) and the observation that Φ_n is a symmetric function of the variables $t_1, t_2, \ldots, t_n$, it follows that $X_j (j = 1, 2, \ldots)$ is a sequence of symmetrically dependent random variables and hence conditionally independent and conditionally equidistributed with respect to the σ-algebra $\mathscr{B}_\infty$ defined as the intersection $\bigcap_{n \ge 1} B_n$, each B_n being the σ-algebra generated by the sets depending symmetrically upon $X_1, \ldots, X_n$ and arbitrarily upon $X_{n+1}, X_{n+2}, \ldots$ [67, p.143]. By [54, p.360] there exists a function $\varphi_x(t)$, which is the Fourier transform of a probability measure on $\mathbb{R}$ for each $x \in \psi$, such that

$$E^{\mathscr{B}_\infty}[e^{itX_j}] = \varphi_x(t) \quad (t \in \mathbb{R}, \, x \in \psi).$$

Now, since the $\{X_j\}_{j=1}^\infty$ are conditionally independent and equidistributed given $\mathscr{B}_\infty$, we have

$$E^{\mathscr{B}_\infty}(e^{i(t_1 X_1 + t_2 X_2 + \cdots + t_n X_n)}) = \prod_{j=1}^n E^{\mathscr{B}_\infty}(e^{it_j X_j})$$

$$= \prod_{j=1}^n \varphi_x(t_j).$$

Upon taking the expectation of both sides we have

$$E(e^{i(t_1 X_1 + t_2 X_2 + \cdots + t_n X_n)}) = E\left(\prod_{j=1}^n \varphi_x(t_j)\right),$$

that is,

$$\Phi_n(t_1, t_2, \ldots, t_n) = E\left(\prod_{j=1}^n \varphi_x(t_j)\right). \tag{7.14}$$

It follows from (7.12) and (7.14) that, for $t \geq 0$,

$$E[|\varphi_x(t) - \varphi_x(-t)|^2] = \Phi_2(t,t) - 2\,\Phi_2(t,-t) + \Phi_2(t,t) = 0.$$

Thus $\varphi_x(t) = \varphi_x(-t)$ for each fixed t with probability one and hence, since the rationals are dense and $\varphi_x(t)$ is continuous in t, $\varphi_x(t) = \varphi_x(-t)$ with probability one for $t \in \mathbb{R}$. Thus, with probability one, $\varphi_x(t)$ is real since it is a Fourier transform.

Since ω is nondecreasing on $\mathbb{R}^+$ there exists a determination of ω^{-1} such that $\omega \circ \omega^{-1}$ is the identity. Choose $t_1, t_2 \geq 0$ and note that

$$E([\varphi_x(\omega^{-1}(t_1 + t_2)) - \varphi_x(\omega^{-1}(t_1))\varphi_x(\omega^{-1}(t_2))]^2) = 0.$$

Hence for any determination of ω^{-1} and fixed $t_1, t_2 \geq 0$,

$$\varphi_x(\omega^{-1}(t_1 + t_2)) = \varphi_x(\omega^{-1}(t_1))\varphi_x(\omega^{-1}(t_2)) \tag{7.15}$$

with probability one. In fact (7.15) holds with probability one for any $t_1, t_2 \geq 0$. This follows from the continuity of $\varphi_x(t)$ in t and the observation that ω^{-1} maps the rationals into a dense subset of the range of ω^{-1} minus possibly a countable set.

It follows from (7.15) that, with probability one,

$$\varphi_x(t) = \exp(-a(x)\omega(t)) \tag{7.16}$$

for all t in the range of ω^{-1}. Since the function $a(x)$ in (7.16) is independent of ω^{-1}, and since one can choose countably many versions of ω^{-1} such that the union of their ranges is dense in $\mathbb{R}^+$, it follows that, with probability one, (7.16) holds for all $t > 0$.

From (7.16), (7.14) and (7.12) we have

$$\int_0^\infty \exp(-u\delta\omega(|t|))\,d\alpha(u) = E(\exp(-a(x)\omega(|t|))) \qquad (t \in R), \tag{7.17}$$

and hence that the random variable $a(x)$ induces the probability distribution $d\alpha(u/\delta)$. If condition (b) holds then the range of $a(x)$ must include values in $(0,\varepsilon)$ for all $\varepsilon > 0$ so, by (7.16), it is possible to choose a sequence a_n of positive numbers tending to 0 such that $\exp(-a_n\,\omega(|t|))$ is positive definite on $\mathbb{R}$. Therefore (Theorem 4.5), $\omega(|s-t|)^{1/2}$ is a function of negative type on $\mathbb{R}$.

In case (a), we define

$$\Phi_n^\lambda(t_1, t_2, \ldots, t_n) = \int_0^\infty \exp\left(-\lambda\delta u \sum_{j=1}^n \omega(|t_j|)\right) d\alpha(u)$$

and argue as above to obtain a nonnegative random variable $a_\lambda(x)$ such that $\exp(-a_\lambda(x)\omega(|t|))$ is positive definite in t for each $x \in \psi$. Then, as in (7.17), the distribution induced by $a_\lambda(x)$ is $d\alpha(u/\lambda\delta)$. And since α is not concentrated at zero there exists a sequence a_n of positive numbers tending to zero such that $\exp(-a_n\,\omega(|t|))$ is again positive definite on $\mathbb{R}$. It follows as before that $\omega(|s-t|)^{1/2}$ is of negative type on $\mathbb{R}$. $\square$

In case $\omega(t) = t^p \ (0 < p < \infty)$ the implication of the preceding argument is that $|s-t|^{p/2}$ is a function of negative type on $\mathbb{R}$ or, equivalently, that $e^{-|t|^p}$ is positive definite on $\mathbb{R}$. But $e^{-|t|^p}$ is not positive definite on R when $2 < p < \infty$ (see Corollary 4.11). Hence the following corollary.

Corollary 7.9. *Let (Ω,μ) be a measure space such that $L^p(\Omega,\mu)$ is infinite-*

dimensional. Then every positive definite and norm-symmetric function Φ *on* L^p
has the form

$$\Phi(f) = \int_0^\infty e^{-u\|f\|_p^p} \, d\alpha(u) \qquad (\|f\|_p > 0) \tag{7.18}$$

for some positive and finite measure α *on* $\mathbb{R}^+$. *Further, if* $2 < p < \infty$, *then* α *must be concentrated at* 0 *so that* Φ *is a constant on* $\|f\|_p > 0$. *If* $0 < p \leq 2$ *and* Φ *is defined by (7.18) with* $\Phi(0) \geq \alpha([0,\infty))$ *then* Φ *is positive definite on* L^p.

Proof. The representation (7.18) follows from Theorem 7.4. Since $L^p(\Omega,\mu)$ is infinite-dimensional, it contains a copy of ℓ^p. This allows us to apply Theorem 7.8 and conclude that α must be concentrated at 0 when $2 < p < \infty$. The last assertion is a consequence of (4.12). $\square$

The continuity assumption at 0 permits the following characterization of RPD(L^p).

Corollary 7.10.   *Let* (Ω,μ) *be a measure space such that* $L^p(\Omega,\mu)$ *is infinite-dimensional. Then, for* $2 < p < \infty$, RPD(L^p) *consists only of constant functions.*

These results suggest that the class of positive definite and norm-symmetric functions on a normed linear space of high linear dimension is likely to consist only of the constants. Such is the case for ℓ^∞, hence also for infinite-dimension L^∞ spaces, and certain algebras of continuous functions. The following lemma is due to Schoenberg [79].

Lemma 7.11.   *Any finite metric space* X_n *of* $n+1$ *points can be embedded in a subspace of* ℓ_n^∞.

Proof. Let $x_0, x_1, \ldots, x_n$ be the points of X_n and let $\overline{x_i x_j}$ be their distances. Consider in ℓ_n^∞ the points

$$y_i = (\overline{x_1 x_i}, \overline{x_2 x_i}, \ldots, \overline{x_n x_i}) \qquad (i = 0, 1, \ldots, n).$$

For their distances in ℓ_n^∞ we have

$$\|y_i - y_j\|_\infty = \max_{1 \leq k \leq n} |\overline{x_k x_i} - \overline{x_k x_j}| = \overline{x_i x_j}.$$

If $i = j = 0$ the equality is obvious as both sides are 0. Otherwise, $|\overline{x_k x_i} - \overline{x_k x_j}| \leq \overline{x_i x_j}$ by the triangular inequality and equality occurs when k equals i or j, whichever one is not 0. $\square$

Theorem 7.12.   *A function* Φ *on* ℓ^∞ *is positive definite and norm-symmetric if and only if there exists numbers* $a \geq b \geq 0$ *such that*

$$\Phi(x) = a \ \text{ if } \ \|x\|_\infty = 0$$

and

$$\Phi(x) = b \ \text{ if } \ \|x\|_\infty > 0.$$

Proof. Since Φ is norm-symmetric the map $\|x\| = \|x\|_\infty \to \Phi(x)$ defines a function f on $\mathbb{R}^+$ such that $f(\|x\|) = \Phi(x)$. The assumption is that

$$\sum_{j,k=1}^N f(\|x_j - x_k\|) \, \xi_j \bar{\xi}_k \geq 0. \tag{7.19}$$

Choose real numbers $0 < r < s \le 2r$ and place on the symbols $p_1, p_2, \ldots, p_N$ the following metric: divide the pairs (i,j), $1 \le i < j \le N$ into two disjoint sets A and B and define

$$\overline{p_i p_j} = r \quad \text{if} \quad (i,j) \in A$$

and

$$\overline{p_i p_j} = s \quad \text{if} \quad (i,j) \in B.$$

According to Lemma 7.11 there exist points $x_1, x_2, \ldots, x_N$ in ℓ^∞ such that

$$\|x_i - x_j\| = r \quad \text{if} \quad (i,j) \in A$$

and

$$\|x_i - x_j\| = s \quad \text{if} \quad (i,j) \in B.$$

With these choices for x_i the inequality (7.19) becomes

$$f(0) \sum_{j=1}^{N} |\xi_j|^2 + 2\left[f(r) \sum_{A} \xi_i \xi_j + f(s) \sum_{B} \xi_i \xi_j \right] \ge 0$$

for real ξ_i. With the $\xi_i = 1$ and B empty this becomes $N f(0) + N(N-1) f(r) \ge 0$, from which it follows, by letting $N \to \infty$, that $f(r) \ge 0$. Next take $N = 2$ with $\xi_1 = 1, \xi_2 = -1$ and B empty to get $f(0) \ge f(r)$. Finally, take $N = 2M$, put $\xi_j = -1$ for $1 \le j \le M$, $\xi_j = 1$ for $M < j \le 2M$, $A = \{(i,j) : \xi_i \xi_j = 1\}$ and $B = \{(i,j) : \xi_i \xi_j = -1\}$. This gives

$$2M f(0) + 2[f(r) \operatorname{Card}(A) - f(s) \operatorname{Card}(B)] \ge 0.$$

Since the ratios $\operatorname{Card}(A)/M^2$ and $\operatorname{Card}(B)/M^2$ tend to one as M increases, this inequality implies that $f(r) \ge f(s)$. Reverse A and B to get $f(s) \ge f(r)$. This shows that f is constant on intervals of the form $[r, 2r]$ and hence constant on $(0, \infty)$. The conclusion for Φ follows by setting $f(0) = a$ and $f(u) = b$ for $0 < u$. This proves the necessity.

A simple exercise involving a regrouping of terms in the quadratic form (7.19) shows that every such Φ does define a norm-symmetric and positive definite function on ℓ^∞. $\square$

Certainly the preceding result, due to Einhorn in $C[0,1]$ [26], carries over to any normed linear space that contains the isometric image of any finite metric space.

Corollary 7.13. *Let X be a topological space and suppose that $C(X)$, the space of bounded complex-valued continuous functions on X, contains n functions having pairwise disjoint support, $n = 1, 2, \ldots$. Then Theorem 7.12 holds with $C(X)$ in place of ℓ^∞.*

Proof. Let n be a positive integer and choose functions $f_1, f_2, \ldots, f_n$ in $C(X)$ of sup-norm 1 and having pairwise disjoint supports. The map ϕ from ℓ_n^∞ to $C(X)$ defined by

$$\phi(x) = \sum_{j=1}^{n} x_j f_j \qquad (x = (x_1, x_2, \ldots, x_n) \in \ell_n^\infty)$$

is clearly an isometry. Therefore, in view of Lemma 7.11, we see that any finite metric space can be embedded in $C(X)$. The proof now follows as before. □

§8. Functions of Negative Type on L^p Spaces

Let (Ω, μ) be a measure space such that $L^p = L^p(\Omega, \mu)$ is infinite-dimensional for $0 < p \le \infty$. Recall that a function F is of negative type on L^p provided that $F: \mathbb{R}^+ \to \mathbb{R}^+$ is continuous, $F(0) = 0$ and $F(\|x - y\|_p)$ is of negative type on L^p. According to Theorem 4.5, $F \in N(L^p)$ if and only if $\exp(-\lambda F^2)$ $(\lambda > 0)$ is radial positive definite on L^p. It then follows from Theorem 7.4 that for every $\lambda > 0$ there exists a finite positive measure α_λ such that $\int_0^\infty d\alpha_\lambda = 1$ and

$$\exp(-\lambda F^2(r)) = \int_0^\infty e^{-r^p u} \, d\alpha_\lambda(u) \qquad (0 \le r < \infty). \tag{8.1}$$

When $2 < p \le \infty$, the right side of (8.1) is a constant function of r in $0 < r < \infty$ by Theorem 7.12 and Corollary 7.10. But the left side is continuous in $0 \le r < \infty$ and $F(0) = 0$; hence $F(r) = 0$ for $0 \le r$.

When $0 < p \le 2$, it is necessary to solve (8.1) for $F(r)$. In anticipation of the forthcoming integral representation, we write

$$\frac{1 - \exp(-\lambda F^2(r))}{\lambda} = \int_0^\infty \frac{1 - e^{-r^p u}}{u} \frac{u}{\lambda} \, d\alpha_\lambda(u)$$

$$= \int_0^\infty \frac{1 - e^{-r^p u}}{u} \, d\beta_\lambda(u)$$

where $d\beta_\lambda(u) = \frac{u}{\lambda} d\alpha_\lambda(u)$. As λ tends to 0 through positive values, the difference quotient on the left converges to $F^2(r)$ uniformly on bounded subsets of $\mathbb{R}^+$. Thus

$$F^2(r) = \lim_{\lambda \downarrow 0} \int_0^\infty \frac{1 - e^{-r^p u}}{u} \, d\beta_\lambda(u). \tag{8.2}$$

The measures $\{\beta_\lambda\}$ need not be uniformly bounded in norm (total variation) or even be finite. Hence a routine weak*-convergence argument will not suffice to calculate the limit on the right in (8.2). Nonetheless we shall show that the $\{\beta_\lambda\}$ do converge in an appropriate sense to a positive measure α (finite on compact sets) on $\mathbb{R}^+$ such that

$$F(r) = \left\{ \int_0^\infty \frac{1 - e^{-r^p u}}{u} \, d\alpha(u) \right\}^{1/2} \qquad (0 \le r) \tag{8.3}$$

and

$$\int_1^\infty \frac{d\alpha(u)}{u} < \infty. \tag{8.4}$$

Theorem 8.1. *The class $N(L^p)\,(0 < p \le \infty)$ contains only the zero function when $2 < p \le \infty$, and when $0 < p \le 2$, it consists of those functions F which have a representation (8.3) with respect to a positive measure on $\mathbb{R}^+$ that satisfies (8.4).*

Proof. Our opening remarks establish the conclusion in case $2 < p \le \infty$.

When $0 < p \leq 2$, $e^{-u\|x-y\|_p^p}$ is positive definite on L^p by (4.12). Hence $F(\|x-y\|_p)$, F being given by (8.3)–(8.4), is of negative type on L^p since it obviously satisfies the condition (2.9). To show that every function in $N(L^p)$ $(0 < p \leq 2)$ has the indicated form we need the following lemma:

Consider the sequence

$$\varphi_n(r) = \int_0^\infty \frac{1-e^{-ru}}{u}\, d\beta_n(u) \qquad (0 \leq r)$$

where β_n is a positive measure on $\mathbb{R}^+$ and

$$\int_1^\infty \frac{d\beta_n(u)}{u} < \infty \qquad (n = 1,2,\ldots).$$

If $\varphi_n(r) \to \varphi(r)$ for each $r \in \mathbb{R}^+$ and φ is continuous at zero, then

$$\varphi(r) = \int_0^\infty \frac{1-e^{-ru}}{u}\, d\beta(u) \qquad (0 \leq r)$$

where β is a positive measure on $\mathbb{R}^+$ such that

$$\int_1^\infty \frac{d\beta(u)}{u} < \infty.$$

The sequence γ_n of positive measures on $[0,\infty]$ defined by $d\gamma_n(u) = \dfrac{1-e^{-u}}{u}\, d\beta_n(u)$ is uniformly bounded since

$$\int_0^\infty d\gamma_n(u) = \varphi_n(1) \to \varphi(1) \quad \text{as} \quad n \to \infty.$$

Hence these measures, considered as elements of the dual space of $C([0,\infty])$, have a weak* limit point γ.

For each $r > 0$,

$$\varphi_n(r) = \int_0^\infty \frac{1-e^{-ru}}{1-e^{-u}}\, d\gamma_n(u) \qquad (n = 1,2,\ldots)$$

and since $\dfrac{1-e^{-ru}}{1-e^{-u}}$ is continuous on $[0,\infty]$, it follows from the definition of weak* convergence that

$$\varphi(r) = \int_0^\infty \frac{1-e^{-ru}}{1-e^{-u}}\, d\gamma(u) \qquad (r \geq 0).$$

Now $\gamma \geq 0$ and $\dfrac{1-e^{-ru}}{1-e^{-u}} \to 1$ as $u \to \infty$; hence $\varphi(r) \geq \gamma(\{\infty\}) \geq 0$ for every $r > 0$. But φ is continuous at 0 and $\varphi(0) = 0$. Hence $\gamma(\{\infty\}) = 0$, proving that γ does not have a point mass at ∞, so we may write

$$\varphi(r) = \int_0^\infty \frac{1-e^{-ru}}{u}\, d\beta(u) \qquad (0 \leq r),$$

where

$$d\beta(u) = \frac{u}{1-e^{-u}}\, d\gamma(u)$$

is clearly a positive measure on $\mathbb{R}^+$ satisfying $\int_1^\infty u^{-1} d\beta(u) < \infty$.

In order to apply the lemma, let $F \in N(L^p)$ $(0 < p \le 2)$, and follow the argument leading to the limit (8.2). In that limit choose a sequence of positive numbers λ_n tending to 0 and replace r by $r^{1/p}$. Then the right side is a sequence which satisfies the conditions of our lemma and its limit is the continuous function $F^2(r^{1/p})$. Now apply the lemma to get the required representation (8.3)–(8.4). This completes the proof. $\square$

This proof from [11] should be compared with the original work of Schoenberg for $p = 2$ [81].

§9. Functions of Negative Type on $\mathbb{R}^N$

It is a consequence of Theorems 4.6 and 6.2 that a function F is of negative type on $\mathbb{R}^N$ if and only if

$$\exp(-\lambda F^2(r)) = \int_0^\infty \Omega_N(ru) \, d\alpha_\lambda(u) \qquad (\lambda > 0, r \ge 0) \tag{9.1}$$

for some finite positive measure α_λ on $\mathbb{R}^+$ of total mass 1. For this class the required limit is

$$F^2(r) = \lim_{\lambda \downarrow 0} \int_0^\infty \frac{1 - \Omega_N(ru)}{u^2} \, d\beta_\lambda(u) \qquad (r \ge 0), \tag{9.2}$$

where $d\beta_\lambda(u) = \dfrac{u^2}{\lambda} \, d\alpha_\lambda(u)$ and the convergence is uniform on compact subsets of $\mathbb{R}^+$.

Theorem 9.1. *The class $N(\mathbb{R}^N)$ $(N = 1, 2, \ldots)$ consists of all functions F which admit a representation*

$$F(r) = \left\{ \int_0^\infty \frac{1 - \Omega_N(ru)}{u^2} \, d\alpha(u) \right\}^{1/2} \qquad (r \ge 0) \tag{9.3}$$

with α a positive measure on $\mathbb{R}^+$ satisfying

$$\int_0^\infty u^{-2} \, d\alpha(u) < \infty. \tag{9.4}$$

Stated in the language of isometric embedding, this remarkable result of Schoenberg and von Nuemann [66] asserts that of the nonnegative and continuous functions F on $\mathbb{R}^+$ with $F(0) = 0$, it is precisely those given by (9.3)–(9.4) for which $\mathbb{R}^N$ with the metric $F(\|x - y\|)$ is embeddable in Hilbert space. We shall present two proofs. The first, following the theme of the previous section, is an outgrowth of the characterization of isometric embedding in terms of quadratic forms given in Theorem 2.4. The second proof utilizes the existence of the embedding and makes full use of the representation for unitary groups of operators.

It is, of course, easy to establish that every F given by (9.3)–(9.4) is a function of negative type on $\mathbb{R}^N$. We need only note that $\Omega_N(ru)$ $(u \ge 0; N = 1, 2, \ldots)$ is radial positive definite on $\mathbb{R}^N$ and hence that $F(\|x - y\|)$ satisfies (2.9). This part of the proof we consider done.

Proof I.   The pattern is similar to the proof of Theorem 8.1. In this case the supporting lemma is the following:
Consider the sequence

$$\varphi_n(r) = \int_0^\infty \frac{1 - \Omega_N(ru)}{u^2} \, d\beta_n(u) \qquad (0 \le r) \tag{9.5}$$

where β_n is a positive measure on $\mathbb{R}^+$ such that

$$\int_1^\infty u^{-2} \, d\beta_n(u) < \infty \qquad (n = 1,2,3,\dots).$$

If $\varphi_n(r) \to \phi(r)$ uniformly on each compact subset of $\mathbb{R}^+$, then

$$\varphi(r) = \int_0^\infty \frac{1 - \Omega_N(ru)}{u^2} \, d\beta(u) \qquad (0 \le r)$$

for some positive measure β on $\mathbb{R}^+$ such that

$$\int_1^\infty u^{-2} \, d\beta(u) < \infty.$$

Suppose first that $N \ge 2$. Then, by (6.4),

$$-1 < \Omega_N(r) < 1 \quad \text{and} \quad \lim_{r \to \infty} \Omega_N(r) = 0. \tag{9.6}$$

The sequence γ_n of positive measures on $[0,\infty]$ defined by $d\gamma_n(u) = \dfrac{1 - \Omega_N(u)}{u^2} \, d\beta_n(u)$ is uniformly bounded since $\int_0^\infty d\gamma_n(u) = \varphi_n(1) \to \phi(1)$ as $n \to \infty$. Let γ be a weak* limit point of the sequence $\{\gamma_n\}$ as elements of the dual of $C([0,\infty])$. For each $r > 0$,

$$\varphi_n(r) = \int_0^\infty \frac{1 - \Omega_N(ru)}{1 - \Omega_N(u)} \, d\gamma_n(u) \qquad (n = 1,2,\dots)$$

and, according to (9.6) and (6.9), the integrand is continuous on $[0,\infty]$. By weak* convergence

$$\varphi(r) = \int_0^\infty \frac{1 - \Omega_N(ru)}{1 - \Omega_N(u)} \, d\gamma(u) \qquad (r \ge 0).$$

Of course $\gamma \ge 0$, and $\dfrac{1 - \Omega_N(ru)}{1 - \Omega_N(u)} \to 1$ as $u \to \infty$ for each $r > 0$; hence $\varphi(r) \ge \gamma(\{\infty\})$ ≥ 0 for $r > 0$ so $0 = \lim_{r \to 0} \phi(r) \ge \gamma(\{\infty\}) \ge 0$. This proves that γ has no mass at ∞ and allows us to write

$$\varphi(r) = \int_0^\infty \frac{1 - \Omega_N(ru)}{u^2} \, d\beta(u) \qquad (r \ge 0),$$

where

$$d\beta(u) = \frac{u^2}{1 - \Omega_N(u)} \, d\gamma(u)$$

is certainly a positive measure on $\mathbb{R}^+$ satisfying $\int_1^\infty u^{-2} d\beta(u) < \infty$.
   This proof does not apply when $N = 1$ for the reason that both assertions of

(9.6) fail for $\Omega_1(r) = \cos r$. However, it can be resurrected by the device of integrating both sides of (9.5) ($N = 1$) from 0 to r obtaining

$$\int_0^r \varphi_n(s)ds = r \int_0^\infty \frac{1-\Omega_3(ru)}{u^2} d\beta_n(u) \qquad (0 \leq r, \, n=1,2,\ldots).$$

Clearly the right side converges uniformly in r on compact subsets of $\mathbb{R}^+$. Hence we may apply the reasoning for the case $N=3$ to conclude that there exists a positive and finite measure γ on $[0,\infty]$ such that

$$\frac{1}{r}\int_0^r \varphi(s)ds = \int_0^\infty \frac{1-\Omega_3(ru)}{1-\Omega_3(u)} d\gamma(u).$$

Further,

$$0 = \varphi(0) = \lim_{r\to 0}\frac{1}{r}\int_0^r \varphi(s)ds \geq \gamma(\{\infty\}) \geq 0.$$

As before, this allows us to write

$$\int_0^r \varphi(s)ds = \int_0^\infty \frac{r-r\Omega_3(ru)}{u^2}d\beta(u), \tag{9.7}$$

where

$$d\beta(u) = \frac{u^2}{1-\Omega_3(u)}d\gamma(u)$$

is a positive measure on $\mathbb{R}^+$ satisfying $\int_1^\infty u^{-2}d\beta(u) < \infty$. The required representation now follows by differentiating both sides of (9.7) with respect to r.

To complete the proof we need only observe that for a sequence of positive λ_n tending to 0, the right side of (9.2) satisfies the assumptions of our lemma and that, therefore, the representation (9.3)–(9.4) is established. $\square$

Proof II.   Let $F \in N(\mathbb{R}^N)$ and let ϕ be an embedding of the F-transform of $\mathbb{R}^N$ into Hilbert space H, that is,

$$\|\phi(s) - \phi(t)\| = F(\|s-t\|) \quad (s\in \mathbb{R}^N, \, t\in \mathbb{R}^N). \tag{9.8}$$

The existence of ϕ is guaranteed by Theorem 4.5. In a rather natural way the isometry ϕ generates a continuous group of unitary operators on a subspace of H.

Let H_1 be the closure of the linear span of the vectors

$$\varphi(t) - \phi(0) \qquad (t\in \mathbb{R}^N)$$

It follows from the continuity of ϕ that H_1 is separable since it is the closed span of linearly independent vectors $\phi(t_j) - \phi(0)$ as t_j ranges over at most a countable subset of $\mathbb{R}^N$. By the Gram-Schmidt process we may orthogonalize this sequence to obtain a finite or countable orthonormal sequence $\{\psi_j\}$ for H_1. Hence there exist real numbers α_{kj} ($j = 1,2,\ldots,k$ and $k = 1,2,\ldots$) such that

$$\psi_k = \sum_{j=1}^k \alpha_{kj}(\phi(t_j)-\phi(0)) \qquad (k=1,2,\ldots). \tag{9.9}$$

Moreover,

$$\phi(t) - \phi(0) = \Sigma_j \, \beta_j(t)\psi_j \qquad (t \in \mathbb{R}^N), \tag{9.10}$$

where $\beta_j(t)$ is a real continuous function of t. From the inner-product relation

$$2(\phi(u+s) - \phi(s), \phi(v+s) - \phi(s)) = \|\phi(u+s) - \phi(s)\|^2 + \|\phi(v+s) - \phi(s)\|^2$$
$$- \|\phi(u+s) - \phi(v+s)\|^2$$

and (9.8), we deduce that

$$(\phi(u+s) - \phi(s), \phi(v+s) - \phi(s)) = (\phi(u) - \phi(0), \phi(v) - \phi(0)) \qquad (u,v,s \in \mathbb{R}^N).$$

And this implies that

$$\|\phi(t+s) - \phi(s) - \Sigma_k \, \beta_k(t) \, [\Sigma_j \, \alpha_{kj}(\phi(t_j+s) - \phi(s))]\|^2 = 0.$$

Therefore, if we introduce the parameter s by defining

$$\psi_k(s) = \sum_{j=1}^{k} \alpha_{kj}(\phi(t_j+s) - \phi(s)) \quad (s \in \mathbb{R}^N), \tag{9.11}$$

it follows that $\{\psi_k(s)\}$ forms an orthonormal system and that

$$\phi(t+s) - \phi(s) = \Sigma_j \, \beta_j(t) \, \psi_j(s) \tag{9.12}$$

for all $s,t \in \mathbb{R}^N$.

By definition H_1 contains the vectors $\phi(t+s) - \phi(0)$, $\phi(s) - \phi(0)$ and hence, by subtraction, $\phi(t+s) - \phi(s)$ for all $s,t \in \mathbb{R}^N$. Thus H_1 contains the closed subspace M_s spanned by the $\{\psi_k(s)\}$, s fixed. According to (9.12) M_s contains all vectors of the form $\phi(t+s) - \phi(s)$, hence $\phi(t) - \phi(s)$ and so also $(\phi(t) - \phi(s)) - (\phi(0) - \phi(s)) = \phi(t) - \phi(0)$. This shows that $M_s = H_1$ for every $s \in \mathbb{R}^N$.

Every $f \in H_1$ has an expansion $f = \Sigma_j \, c_j \, \psi_j(0)$ with real c_j satisfying $\Sigma c_j^2 < \infty$. For each $s \in \mathbb{R}^N$ define an operator $U(s)$ by

$$U(s)f = \sum_j c_j \, \psi_j(s).$$

This operator is unitary and, by (9.12),

$$U(s)\,(\phi(t) - \phi(0)) = \Sigma_j \, \beta_j(t) \, \psi_j(s) = \phi(t+s) - \phi(t) \tag{9.13}$$

for all $s,t \in \mathbb{R}^N$. In particular, if we put $t = u$ and $t = v$ and subtract this gives

$$U(s)\,(\phi(u) - \phi(v)) = \phi(u+s) - \phi(v+s).$$

From this, letting s be s_1, s_2 and $s_1 + s_2$ we conclude that

$$U(s_1)U(s_2) = U(s_1 + s_2) \qquad (s_1 \in \mathbb{R}^N, \, s_2 \in \mathbb{R}^N). \tag{9.14}$$

Finally, from the continuity of each $\psi_k(s)$ in s and the definition of $U(s)f$, we conclude that

$$\|U(s)f - f\| \to 0 \quad \text{as} \quad \|s\| \to 0 \qquad (f \in H_1). \tag{9.15}$$

From (9.14) and (9.15) it follows that $U(s)$ $(s \in \mathbb{R}^N)$ is a strongly continuous group of unitary operators on H_1.

Let H' be the complexification of H and H_1' the complexification of H_1. Then each $U(s)$ extends to a unitary operator, $U(s)$ extends to a strongly continuous unitary group of operators on H_1' and (9.8) and (9.13) remain unchanged. Thus we can apply Stone's theorem to conclude the existence of a resolution of the identity E on the Borel subsets of $\mathbb{R}^N$ and a densely defined self-adjoint operator A on H_1' such that

$$A = \int_{\mathbb{R}^N} x \, dE(x) \quad \text{and} \quad U(s) = \exp(isA) = \int_{\mathbb{R}^N} e^{i(s,x)} \, dE(x) \quad (s \in \mathbb{R}^N).$$

Choose $0 < \delta < 1$ and let $I_1, I_2, I_3, \ldots, I_m$ be a partition of S_{N-1} into disjoint Borel sets, each with nonempty interior if $N > 1$, and such that $|(u,v)| \geq \delta$ for all $u, v \in I_j$ for $j = 1, 2, \ldots, m$. Then partition $\mathbb{R}^N$ into the sets $I_\infty = \{0\}$ and

$$I_{jk} = \{x \in \mathbb{R}^N : x/\|x\| \in I_j \text{ and } 2^{-k-1} < \|x\| \leq 2^{-k}\} \quad (j = 1, \ldots, m, \ k = 0, \pm 1, \pm 2, \ldots).$$

Let M_{jk} be the range of the self-adjoint projection $P_{jk} = E(I_{jk})$, M_∞ the range of $P_\infty = E(I_\infty)$ and put $\phi_\infty(t) = P_\infty \phi(t)$ and $\phi_{jk}(t) = P_{jk} \phi(t)$. Since the subspaces M_∞ and M_{jk} $(j = 1, \ldots, m; \ k = 0, \pm 1, \pm 2, \ldots)$ are mutually orthogonal and span H_1', the vector $\phi(t) - \phi(0)$ has the decomposition

$$\phi(t) - \phi(0) = \phi_\infty(t) - \phi_\infty(0) + \sum_{k=-\infty}^{\infty} \sum_{j=1}^{m} (\phi_{jk}(t) - \phi_{jk}(0)) \quad (t \in \mathbb{R}^N).$$

It follows from (9.8) that

$$F(\|t\|)^2 = \|\phi_\infty(t) - \phi_\infty(0)\|^2 + \sum_{k=-\infty}^{\infty} \sum_{j=1}^{m} \|\phi_{jk}(t) - \phi_{jk}(0)\|^2 \quad (t \in \mathbb{R}^N). \quad (9.16)$$

The remainder of the proof is devoted to showing that this series can be written in the form (9.3)–(9.4).

Since M_∞ and each M_{jk} is a reducing subspace for $\exp(isA)$, it follows from (9.13) that

$$\phi_{jk}(t+s) = \exp(isA)(\phi_{jk}(t) - \phi_{jk}(0)) + \phi_{jk}(s) \quad (9.17)$$

and

$$\phi_\infty(t+s) = \exp(isA)(\phi_\infty(t) - \phi_\infty(0)) + \phi_\infty(s).$$

But $\exp(isA) = \text{id.}$ on M_∞ so that

$$\phi_\infty(t+s) - \phi_\infty(0) = (\phi_\infty(t) - \phi_\infty(0)) + (\phi_\infty(s) - \phi_\infty(0))$$

for all s and t in $\mathbb{R}^N$. By continuity there exist points $w_1, w_2, \ldots, w_n$ in M_∞ such that

$$\phi_\infty(t) - \phi_\infty(0) = \sum_{j=1}^{N} t_j w_j \quad \text{where} \quad t = (t_1, t_2, \ldots, t_N). \quad (9.18)$$

When $u \in M_{jk}$ and $s \in \mathbb{R}^N$,

$$\exp(isA)u = \exp(isA)P_{jk}u = \int_{\mathbb{R}^N} e^{i(s,x)} \, dE(x) P_{jk}u = \int_{I_{jk}} e^{i(s,x)} \, dE(x)u.$$

Hence $1 - \exp(isA)$ is invertible on M_{jk} provided that $1 - e^{i(s,x)}$ is bounded away from zero on I_{jk}, and this is the case provided that s is in the set

$$K_{jk} = \{s : \pm s/\|s\| \in I_j \text{ and } 0 < \|s\| < 2^k \pi\}. \quad (9.19)$$

Thus $(1-\exp(isA))^{-1}$ exists on M_{jk} provided that s is in K_{jk}. The relation

$$(\exp(itA)-1)(\phi_{jk}(s)-\phi_{jk}(0)) = (\exp(isA)-1)(\phi_{jk}(t)-\phi_{jk}(0))$$

results from interchanging s and t in (9.17) and subtracting. Hence

$$(\exp(isA)-1)^{-1}(\phi_{jk}(s)-\phi_{jk}(0)) = (\exp(itA)-1)^{-1}(\phi_{jk}(t)-\phi_{jk}(0))$$

provided both s and t are in K_{jk}. Since the vector

$$v_{jk} = (\exp(isA)-1)^{-1}(\phi_{jk}(s)-\phi_{jk}(0))$$

is independent of s in K_{jk}, the formula

$$\phi_{jk}(s) = (\exp(isA)-1)v_{jk} + \phi_{jk}(0) \tag{9.20}$$

holds for all s such that $0 < \|s\| < 2^k\pi$ and $\pm s/\|s\| \in I_j$, since it is obvious when $s=0$. Further, substitution of (9.20) into (9.17) shows that if (9.20) holds for s and t, it also holds for $s+t$. Hence (9.20) holds for all s in the cone

$$I_j = \bigcup_{k=-\infty}^{\infty} I_{jk},$$

its negative $-I_j$ and $s=0$. Since I_j has a nonempty interior, $C_j = I_j \cup [-I_j] \cup \{0\}$ is such that

$$\mathbb{R}^N = \{s+t : s \in C_j,\ t \in C_j\}.$$

Consequently, (9.20) is valid for all $s \in \mathbb{R}^N$. Thus we have

$$\|\phi_{jk}(s)-\phi_{jk}(0)\|^2 = \|(\exp(isA)-1)v_{jk}\|^2 = \int_{I_{jk}} \|e^{i\langle s,x\rangle}-1\|^2\, d\|E(x)v_{jk}\|^2$$

for every $s \in \mathbb{R}^N$.

Now define a Borel measure v on $\mathbb{R}^N \setminus \{0\}$ by setting $v(\cdot) = \|E(\cdot)v_{jk}\|^2$ on I_{jk} for $j=1,2,3,\ldots,m$ and $k=0,\pm 1,\pm 2,\ldots$. This allows us to write (9.16) in the form

$$F(\|t\|)^2 = \left\|\sum_1^N t_j w_j\right\|^2 + 2\int_{\mathbb{R}^N\setminus\{0\}} (1-\cos(t,x))dv(x). \tag{9.21}$$

In this expression make the substitution $t=r\xi$, where $r=\|t\|$ and $\xi \in S_{N-1}$, then integrate both sides over S_{N-1} with respect to $\omega_{N-1}^{-1} d\sigma_{N-1}(\xi)$. By (6.4) the result is

$$F(r)^2 = r^2 w_0 + 2\int_{\mathbb{R}^N\setminus\{0\}} [1-\Omega_N(r\|x\|)]dv(x).$$

Just as in the proof of Theorem 5.1 this integral may be reduced from an integral over $\mathbb{R}^N \setminus \{0\}$ to an integral over $(0,\infty)$ with respect to some positive measure β:

$$F(r)^2 = r^2 w_0 + 2\int_{0^+}^{\infty} [1-\Omega_N(ru)]d\beta. \tag{9.22}$$

The measure β is bounded at ∞, that is, $\int_a^{\infty} d\beta < \infty$ for every $a > 0$. When $N \geq 2$ this is a consequence of the fact that $\Omega_N(ru) \to 0$ as $u \to \infty$ for $r > 0$. When $N=1$, integrate both sides of (9.22) and apply the same reasoning. This leaves the growth of β near 0 in doubt. However, if we can show that

$$\int_{(0,1)} u^2\, d\beta(u) < \infty, \tag{9.23}$$

then the positive measure α defined by $\alpha(\{0\}) = w_0$ and $\alpha(A) = 2 \int_A u^2 d\beta(u)$ for every Borel subset A of $(0,\infty)$ will certainly satisfy (9.4) and (9.22) will reduce to (9.3).

From (9.6) and the series expansion (6.9) when $N > 1$, and from special properties of $\cos(u)$ when $N = 1$, we see that $u^{-2}(1 - \Omega_N(u)) \geq \delta_N > 0$ for $0 < u \leq 1$. Hence

$$F^2(1) \geq w_0 + 2\,\delta_N \int_{(0,1)} u^2 d\beta(u)$$

and (9.23) follows. The proof is complete. $\square$

In the important special case $N = 1$, this result takes the following form.

Corollary 9.2. *A function F belongs to* $N(\mathbb{R})$ *if and only if*

$$F(r) = \left(\int_0^\infty \frac{\sin^2(ru)}{u^2} d\alpha(u) \right)^{1/2} \qquad (0 \leq r) \tag{9.24}$$

for some positive measure on $\mathbb{R}^+$ *which satisfies (9.4).*

If α is a sequence of point masses $A_1 u_1{}^2, A_2 u_2{}^2, \ldots, A_k u_k{}^2$ at $0 < u_1 < u_2 < \ldots < u_k$ and $C = \alpha(\{0\})$, then

$$F^2(r) = Cr^2 + \sum_{j=1}^k A_j \sin^2(ru_j).$$

An F-transform of $\mathbb{R}$ of this type was considered in Example 3.4. In that case α consisted of point masses at 0 and 1, and the resulting embedding was into $\mathbb{R}^3$.

Theorem 9.3. *Let* $F \in N(\mathbb{R}^1)$. *Then the F-transform of* $\mathbb{R}^1$ *is embeddable in Euclidean space if and only if the measure α consists of a finite number k of point masses. Specifically, the embedding lies in* $\mathbb{R}^N$ *and in no proper subspace of* $\mathbb{R}^N$ *if and only if*

$$C = 0, \ N = 2k \qquad \text{for } N \text{ even,} \tag{9.25}$$

$$C > 0, \ N = 2k - 1 \qquad \text{for } N \text{ odd.} \tag{9.26}$$

Proof. In case $C = 0$, it is easy to verify that

$$\phi(t) = (\tfrac{1}{2}A_1^{1/2}\cos 2u_1 t, \ldots, \tfrac{1}{2}A_k^{1/2}\cos 2u_k t, \tfrac{1}{2}A_1^{1/2}\sin 2u_1 t, \ldots, \tfrac{1}{2}A_k^{1/2}\sin 2u_k t) \ (t \in \mathbb{R})$$

defines an embedding of $F(R)$ into $\mathbb{R}^{2k}$. If $C > 0$, the embedding is

$$\phi(t) = (Ct, \tfrac{1}{2}A_1^{1/2}\cos 2u_1 t, \ldots, \tfrac{1}{2}A_{k-1}\cos 2u_{k-1} t, \tfrac{1}{2}A_1^{1/2}\sin 2u_1 t, \ldots, \tfrac{1}{2}A_{k-1}^{1/2}\sin 2u_{k-1} t)$$
$$(t \in \mathbb{R}).$$

The embeddings lie in $\mathbb{R}^{2k}$ and $\mathbb{R}^{2k-1}$, respectively, but in no lower dimensional subspace.

Now suppose that α does not consist exclusively of a finite number of point masses. Choose $n + 1$ distinct numbers $t_0(=0), t_1, t_2, \ldots, t_n$ and $\xi_1, \xi_2, \ldots, \xi_n$. It follows from the identity (3.4) that

$$\sum_{j,k=1}^n (F^2(t_j) + F^2(t_k) - F^2(t_j - t_k))\xi_j \xi_k$$

$$= \int_0^\infty \left\{ 2\left(\sum_{j=1}^n \xi_j \sin^2 t_j u \right)^2 + \frac{1}{2}\left(\sum_{j=1}^n \xi_j \sin 2t_j u \right)^2 \right\} u^{-2} d\alpha(u) \geq 0.$$

Were this equal to 0 it would follow that $\sum_{j=1}^n \xi_j \sin 2t_j u = 0$ for all u in the support of α, hence in an infinite set; but this is impossible unless $\sum \xi_j^2 = 0$. In other words, the quadratic form is positive definite which, according to Corollary 2.2, means that the points $t_0(=0), t_1, t_2, \ldots, t_n$ can be embedded in R^n but not in R^r for $r < n$. As n can be made indefinitely large, the range of the embedding cannot lie in any Euclidean space. $\square$

In the same setting it is clear that the range of an embedding in Hilbert space is a bounded set if and only if the function F is bounded. The following result provides conditions which insure the boundedness of a function in $N(\mathbb{R})$.

Theorem 9.4. *A function F in $N(\mathbb{R})$ is bounded if and only if*

$$\alpha(\{0\}) = 0 \ \text{ and } \ \int_{0+}^\infty u^{-2} d\alpha(u) < \infty. \tag{9.27}$$

Proof. Suppose F is bounded. It is clear from (9.24) that $F^2(r) \geq r^2 \alpha(\{0\})$ for $r > 0$; hence $\alpha(\{0\}) = 0$. Next let

$$F_{\varepsilon, a}(t) = \int_\varepsilon^a \frac{\sin^2 tu}{u^2} d\alpha(u) \qquad (0 < \varepsilon < a < \infty).$$

The mean-value of $F_{\varepsilon, a}$ over the interval $[0, T]$ may be written as

$$\frac{1}{T} \int_0^T F_{\varepsilon, a}(t) dt = \int_0^T \left(\frac{1}{T} \int_\varepsilon^a \frac{\sin^2 tu}{u^2} d\alpha(u) \right) dt$$

$$= \int_\varepsilon^a \left(\frac{1}{2} - \frac{\sin 2Tu}{4Tu} \right) u^{-2} d\alpha(t)$$

and hence

$$\lim_{T \to \infty} \frac{1}{T} \int_0^T F_{\varepsilon, a}(t) dt = \frac{1}{2} \int_\varepsilon^a u^{-2} d\alpha(t).$$

Since $F(t) \geq F_{\varepsilon, a}(t)$ we have

$$\overline{\lim_{t \to \infty}} F(t) \geq \overline{\lim_{t \to \infty}} F_{\varepsilon, a}(t) \geq \frac{1}{2} \int_\varepsilon^a u^{-2} d\alpha(t).$$

Thus the integral is bounded above independent of ε and a. The second part of (9.27) now follows by allowing $\varepsilon \to 0$ and $a \to \infty$. The converse is obvious. $\square$

We state without proof the following additional result. For proofs, see [66].

Theorem 9.5. *Let ϕ be an embedding of $F(\mathbb{R})$ into H. Then the curve $\Gamma = \{\phi(t) : t \in \mathbb{R}\}$ is rectifiable if and only if α is bounded; it is a closed curve if and only if F is periodic and has the form $F^2(t) = \sum_{j=1}^\infty c_j \sin^2\left(\frac{j\pi}{\tau} t\right)$, $\tau > 0$, $c_j \geq 0$, $\sum_1^\infty c_j < \infty$; and it is bounded if and only if it can be placed on a sphere.*

Chapter III. The Extension Problem for Contractions and Isometries

§10. Formulation

Let (X,d_1) and (Y,d_2) be metric spaces. A mapping T from a subset S of X into Y is a *contraction* provided that

$$d_2(Tx_1,Tx_2) \leq d_1(x_1,x_2) \qquad (x_1,x_2 \in S), \tag{10.1}$$

an *isometry* if equality is maintained for all pairs, or a *Lipschitz-Hölder* map of order α (a Lip(α) map) with constant k provided that

$$d_2(Tx_1,Tx_2) \leq k\, d_1(x_1,x_2)^\alpha \qquad (x_1,x_2 \in S). \tag{10.2}$$

In this chapter we consider the problem of determining conditions under which contractions, isometries, or Lip(α) maps from various subsets of X into Y can always be extended to a map of the same type (with preservation of the constant for Lip(α) maps) from X into Y. In most instances Y will be a normed linear space, and in such cases it is no restriction to take $k = 1$ in (10.2). This we will do. We shall also make the further restriction $0 < \alpha \leq 1$ in (10.2), thereby making it possible to view a Lip(α) map as a contraction on X with the transformed metric d_1^α. This is a natural restriction on α since the map $2 \to 2^{2\alpha-1}$, $-2 \to 2^{2\alpha-1}$ is Lip(α) but cannot be extended to 0 when $\alpha > 1$.

Definition 10.1. *The pair (X,Y) is said to have the contraction (isometric) extension property if corresponding to every contraction (isometry) T from an arbitrary subset S of X into Y, there exists a contraction (isometry) T of X into Y such that $\tilde{T}|_s = T$, that is, the restriction of $\tilde{T}$ to S is T.*

The requirement that contractions from X into Y always extend is severe and cannot be met in general. A simple example will serve to illustrate this point.

Example 10.2. In $\mathbb{R}^2$ let H be a regular hexagon with center at the origin and C a circle inscribed in H. Let $\|\cdot\|_H$ and $\|\cdot\|_C$ be the two norms on $\mathbb{R}^2$ whose unit circles are H and C respectively. Suppose x_1 and x_2 are two consecutive points of contact between H and C and let $S = \{0,x_1,x_2\}$. Define T on S by $T(0) = 0$, $T(x_1) = x_1$ and $T(x_2) = x_2$. Since

$$\|x_1\|_H = \|x_1\|_C = \|x_2\|_H = \|x_2\|_C = \|x_1 - x_2\|_H = \|x_1 - x_2\|_C = 1,$$

T is a contraction from $(\mathbb{R}^2, \|\cdot\|_H)$ into $(\mathbb{R}^2, \|\cdot\|_C)$. However T cannot be extended to $x = (x_1 + x_2)/3$ because $\|x\|_H = \|x - x_1\|_H = \|x - x_2\|_H = \frac{1}{2}$ and any choice $Tx = y$ would have to satisfy $\|y\|_C \leq \frac{1}{2}$, $\|y - x_1\|_C \leq \frac{1}{2}$ and $\|y - x_2\|_C \leq \frac{1}{2}$. Such a choice of Y is clearly impossible.

The geometric relationship imposed on metric spaces which have the contraction extension property is best understood through the following concept concerning families of closed balls. Throughout this chapter $B(x,r)$ will denote the closed ball about x of radius r.

Definition 10.3. *The pair (X,Y) has property (K) provided that whenever $\{B(x_i,r_i) : i \in I\}$ and $\{B(y_i,r_i) : i \in I\}$ are two families of balls in X and Y respectively, each indexed over a set I, such that*

$$d_2(y_i,y_j) \leq d_1(x_i,x_j) \qquad (i,j \in I), \tag{10.3}$$

then

$$\bigcap_{i \in I} B(x_i,r_i) \neq \emptyset \tag{10.4}$$

implies

$$\bigcap_{i \in I} B(y_i,r_i) \neq \emptyset. \tag{10.5}$$

M.D. Kirszbraun [44] first established this property for $(\mathbb{R}^n, \mathbb{R}^n)$ and, as we will now show, it is fundamental to the contraction extension problem. Also see Valentine [86].

§11. The Kirszbraun Intersection Property

Theorem 11.1. *The metric pair (X,Y) has the contraction extension property if and only if (X,Y) has property (K).*

Proof. First suppose (X,Y) has the contraction extension property and two families of balls are given in X and Y satisfying (10.3) and (10.4). Let $x \in \bigcap_{i \in I} B(x_i,r_i)$. The map T defined by $T(x_i) = y_i$ $(i \in I)$ is, by (10.3), a contraction from $S = \{x_i : i \in I\}$ into Y. By assumption T can be extended to all of X and, in particular, to x. Thus there exists $y \in Y$ such that

$$d_2(y,y_i) \leq d_1(x,x_i) \leq r_i \qquad (i \in I)$$

and, therefore, $y \in \bigcap_{i \in I} B(y_i,r_i)$ and (10.5) is satisfied.

For the converse, let $T : S \to Y$ be a contraction from the proper subset S of X. We extend T point-by-point. Choose $x \in X \setminus S$. The collections of balls $\{B(w, d_1(x,w)) : w \in S\}$ and $\{B(T(w), d_1(x,w)) : w \in S\}$ satisfy (10.3) because T is a contraction. Since $x \in \bigcap_{w \in S} B(w,d_1(x,w))$ and (K) holds, there exists a $y \in \bigcap_{w \in S} B(Tw, d_1(x,w))$. Setting $Tx = y$ clearly extends T as a contraction to $S \cup \{x\}$. When X is separable one can inductively extend T to a dense subset of X and then to all of X by continuity.

In general we proceed by partially ordering the family $\mathcal{E}$ of all contractions from subsets of X into Y which extend T. For $T_1, T_2 \in \mathcal{E}$, define $T_1 \leq T_2$ provided $\mathcal{D}(T_1) \subseteq \mathcal{D}(T_2)$ and T_2 extends T_1. Every totally ordered subfamily $\mathcal{E}'$ of $\mathcal{E}$ has a maximal element $\tilde{T}$ with $\mathcal{D}(\tilde{T}) = \bigcup_{T' \in \mathcal{E}'} \mathcal{D}(\tilde{T}')$ and

$$\tilde{T}(x) = T'(x) \text{ for } T' \in \mathscr{E}' \text{ and } x \in \mathscr{D}(T').$$

By Zorn's lemma, $\mathscr{E}$ contains a maximal element $\tilde{T}$. Clearly $\tilde{T}$ extends T and has domain X, for in the contrary case we could extend the domain to one more point, contradicting the maximality of $\tilde{T}$. $\square$

A metric space Y is said to have the *binary intersection property* if every collection of mutually intersecting balls in Y has nonempty intersection. The space Y is *metrically convex* if $x, y \in Y$ and $0 < \lambda < 1$ imply the existence of $z \in Y$ with $d_2(x,z) = \lambda d_2(x,y)$ and $d_2(y,z) = (1-\lambda)d_2(x,y)$. These are precisely the spaces for which (X,Y) has the contraction extension property for every metric space X.

Theorem 11.2. *If Y is a metric space, then (X,Y) has the contraction extension property for every metric space X if and only if Y is metrically convex and has the binary intersection property. In this case (X,Y) has the $\mathrm{Lip}(\alpha)$ $(0 < \alpha \leq 1)$ extension property for every metric space X.*

Proof. Suppose Y is metrically convex and has the binary intersection property. It suffices, by the introductory remarks and Theorem 11.1, to show that (X,Y) has property (K) for every metric space X. Let X be a metric space and $\{B(x_i, r_i) : i \in I\}$ and $\{B(y_i, r_i) : i \in I\}$ collections of balls in X and Y satisfying

(10.3) and (10.4). Since $\bigcap_{i \in I} B(x_i, r_i) \neq \emptyset$, it follows that $d_1(x_i, x_j) \leq r_i + r_j$ $(i,j \in I)$.

Then $d_2(y_i, y_j) \leq r_i + r_j$ $(i,j \in I)$ and since Y is metrically convex, the collection $\{B(y_i, r_i) : i \in I\}$ is mutually intersecting. From the binary intersection property it

follows that $\bigcap_{i \in I} B(y_i, r_i) \neq \emptyset$ and Y has property (K).

Conversely, suppose (X,Y) has the contraction extension property for every metric space X. If $x, y \in Y$, let $z, w \in R$ so that $|z - w| = d_2(x,y)$. The map $T(z) = x$ and $T(w) = y$ is a contraction and can be extended to $\lambda z + (1 - \lambda)w$ $(0 < \lambda < 1)$. It follows that Y is metrically convex.

Let $\{B(y_i, r_i) : i \in I\}$ be a collection of mutually intersecting balls in Y. Let $X = I \cup \{I\}$ with metric d_1 defined by $d_1(i,j) = r_i + r_j$ $(i,j \in I)$ and $d_1(i, \{I\}) = r_i$ $(i \in I)$. Obviously d_1 is a metric on X. For $i, j \in I$, $i \neq j$, $B(y_i, r_i) \cap B(y_j, r_j) \neq \emptyset$, so $d_2(y_i, y_j) \leq r_i + r_j$. Therefore the collections $\{B(i, r_i) : i \in I\}$ and $\{B(y_i, r_i) : i \in I\}$ satisfy (10.3)

and (10.4). By assumption (X,Y) has property (K) and thus $\bigcap_{i \in I} B(y_i, r_i) \neq \emptyset$. $\square$

From the work of Nachbin [65] and Kelley [43] on spaces with the Hahn-Banach extension property, it follows that the Banach spaces with the binary intersection property are precisely those of the form $C(K)$, where K is an extremally disconnected compact Hausdorff space. It is easy to see that ℓ_n^∞ with the supremum norm has the binary intersection property. In §12 we give conditions sufficient to imply the binary intersection property.

Theorem 11.3. *If H is a Hilbert space, then (H,H) has the extension property for contractions and $\mathrm{Lip}(\alpha)$ maps, $0 < \alpha \leq 1$.*

Proof. By Theorem 11.1, this is equivalent to showing that (X,Y) has property (K), where X is H with the transformed metric $\|x-y\|^\alpha$ $(0 < \alpha \leq 1)$ and Y is H with the usual norm $\|x-y\|$.

Let $\{B(x_i, r_i) : i \in I\}$ and $\{B(y_i, r_i) : i \in I\}$ be collections of balls in X and Y respectively, satisfying (10.3) and (10.4). Pick an index $0 \in I$ and define

$$B_i = B(y_0, r_0) \cap B(y_i, r_i) \qquad (i \in I).$$

Each of the weakly closed sets B_i is contained in the weakly compact ball $B(y_0, r_0)$. Thus (10.5) will follow provided $\{B(y_i, r_i) : i \in I\}$ possesses the finite intersection property.

To this end choose finitely many balls indexed $i = 1, \ldots, m$. Conditions (10.3) and (10.4) become

$$\|y_i - y_j\| \le \|x_i - x_j\|^\alpha \qquad (1 \le i, j \le m), \tag{11.1}$$

and for some $x \in X$,

$$\|x - x_i\|^\alpha \le r_i \qquad (1 \le i \le m). \tag{11.2}$$

We may assume $x \ne x_i$ ($1 \le i \le m$) for if $x = x_j$ then $y = y_j \in \bigcap_{i=1}^{m} B(y_i, r_i)$.

Define the real function f on H by

$$f(t) = \max_{1 \le i \le m} \frac{\|t - y_i\|}{\|x - x_i\|^\alpha}.$$

Since f is continuous, $\lim_{\|t\| \to \infty} f(t) = +\infty$ and $f(t) \ge f(t_p)$, where t_p is the orthogonal projection of t onto the linear span L of $\{y_1, \ldots, y_m\}$. So f assumes its minimum value λ at some point $y \in L$. The proof is complete provided $\lambda \le 1$, so assume $\lambda > 1$ and, by relabeling points, that

$$\|y - y_i\| = \lambda \|x - x_i\|^\alpha \quad \text{for } i = 1, \ldots, k \tag{11.3}$$

and

$$\|y - y_i\| < \lambda \|x - x_i\|^\alpha \quad \text{for } i = k+1, \ldots, m. \tag{11.4}$$

It follows that y is a member of the convex hull C of $\{y_1, \ldots, y_k\}$; otherwise there would exist a hyperplane Π in L separating y from C and a small displacement of y toward Π would reduce the left members of (11.3) and preserve the inequalities in (11.4) contrary to the minimality of λ. Hence there exist nonnegative numbers $\xi_1, \ldots, \xi_k$ with $\Sigma_{i=1}^k \xi_k = 1$ and such that

$$y = \sum_{i=1}^{k} \xi_i \, y_i. \tag{11.5}$$

Expanding the left member of

$$\|y_i - y_j\|^2 \le \|x_i - x_j\|^{2\alpha} \qquad (1 \le i, j \le k)$$

we obtain

$$\|y_i - y\|^2 - 2\mathrm{Re}(y_i - y, y_j - y) + \|y_j - y\|^2 \le \|x_i - x_j\|^{2\alpha} \qquad (1 \le i, j \le k). \tag{11.6}$$

Combining (11.3) and (11.6) we find

$$\sum_{i,j=1}^{k} \xi_i \xi_j \{\|x-x_i\|^{2\alpha} + \|x-x_j\|^{2\alpha}\} < \sum_{i,j=1}^{k} \xi_i \xi_j \|x_i-x_j\|^{2\alpha}$$

$$+ \sum_{i,j=1}^{k} \xi_i \xi_j 2\operatorname{Re}(y_i-y, y_j-y).$$

The second sum on the right is 0 by (11.5) and therefore

$$2 \sum_{i=1}^{k} \xi_i \|x-x_i\|^{2\alpha} < \sum_{i,j=1}^{k} \xi_i \xi_j \|x_i-x_j\|^{2\alpha},$$

contrary to the inequality (4.15) in the case $p = 2$. Thus $\lambda \leq 1$ and $\bigcap_{i=1}^{m} B(y_i,r_i) \neq \emptyset$. □

For the origins of this proof see [82] and [59].

Theorem 11.4. *If H is a Hilbert space then (H,H) has the isometric extension property if and only if H is finite-dimensional. In general, if $S \subseteq H$ and $T : S \to H$ is an isometry, then T can be extended as an isometry to the closed linear span of S.*

Proof. Suppose $S \subseteq H$ and $T : S \to H$ is an isometry. Let t_b denote translation by a fixed vector b, that is $t_b(x) = x+b$ $(x \in H)$. Fix $a \in S$ and set $S' = S-a$ and $b = -Ta$. Then $T' = t_b T t_a$ is an isometry from S' into H with $T'(0) = 0$. Therefore $\|T'x\| = \|x\|$ and $(T'x,T'y) = (x,y)$ for $x,y \in S'$. Define $\tilde{T}'$ on the linear span V of S' by

$$\tilde{T}'\left(\sum_{j=1}^{n} a_j x_j\right) = \sum_{j=1}^{n} a_j T' x_j, \quad x_1,\ldots,x_n \in S \text{ and } a_1,\ldots,a_n \text{ real.}$$

The fact that T' is a well-defined isometry follows from the identity

$$\left\|\sum_{j=1}^{n} a_j T' x_j\right\|^2 = \sum_{j,k=1}^{n} a_j a_k (T'x_j,T'x_k) = \sum_{j,k=1}^{n} a_j a_k (x_j,x_k) = \left\|\sum_{j=1}^{n} a_j x_j\right\|^2.$$

Thus $\tilde{T}'$ is a linear isometry from V onto the linear span W of $T'(S')$, and $t_{-b} \tilde{T}' t_{-a}$ is an isometry on the linear span of S that extends T. If H is finite-dimensional, V and W have the same linear dimension and so do their orthogonal complements. Hence there is a linear isometry from the orthogonal complement of V onto the orthogonal complement of W giving rise, in an obvious way, to an isometry T'' of H onto H that extends T'. Consequently $t_{-b} T'' t_{-a}$ is an isometry that extends T.

A simple example in the sequence space ℓ^2 will illustrate the necessity of finite-dimensionality for the isometric extension property. Let $S = \{x = (x_1,x_2,\ldots) \in \ell^2 : x_1 = 0\}$ and define $T : S \to \ell^2$ by $T(x) = (x_2,x_3,\ldots)$. Clearly T is an isometry which cannot be extended to the vector $(1,0,0,\ldots)$. □

As the following counterexample [85] shows, it is not in general possible to extend contractions in classical Banach spaces which are not Hilbert spaces.

Theorem 11.5. *Let ℓ_n^p, $1 < p < \infty$, $n \geq 1$, be the n-dimensional Euclidean space with norm*

$$\|x\|_p = \|(x_1,\dots,x_n)\|_p = \left(\sum_{i=1}^{n} |x_i|^p\right)^{1/p}.$$

Then if $n > 1$ and $p \neq 2$, (ℓ_n^p, ℓ_n^p) does not have property (K).

Proof. Consider the points $x_1 = (0,0,\dots,0)$, $x_2 = (0,1,0,\dots,0)$ and $x_3 = (1,0,\dots,0)$ and the collection of balls $\{B_i = B(x_i, 2^{(1-p)/p})\}_{i=1}^3$. Since $\|x_1 - x_2\|_p = \|x_1 - x_3\|_p = 1$ and $\|x_2 - x_3\|_p = 2^{1/p}$, we have

$$\bigcap_{i=1}^{3} B_i = \left\{\left(\tfrac{1}{2},\tfrac{1}{2},0,\dots,0\right)\right\}.$$

Now let $y_1 = x_1$, $y_2 = ((1-2^{1-p})^{1/p}, 2^{(1-p)/p}, 0, \dots, 0)$ and $y_3 = ((1-2^{1-p})^{1/p}, -2^{(1-p)/p}, 0, \dots, 0)$ and consider the collection $\{B_i' = B(y_i, 2^{(1-p)/p})\}_{i=1}^3$. Since $\|y_1 - y_2\|_p = \|y_1 - y_3\|_p = 1$ and $\|y_2 - y_3\|_p = 2^{1/p}$, the collections satisfy (10.3) and (10.4). However if $p > 2$, $\bigcap_{i=1}^{3} B_i' = \emptyset$. Uniform convexity shows that $B_2' \cap B_3' = \{((1-2^{1-p})^{1/p}, 0, \dots, 0)\} = \{y\}$ and $y \notin B_1'$ since $(1-2^{1-p})^{1/p} > 2^{(1-p)/p}$. In the case $1 < p < 2$ we need only reverse the roles of x_1, x_2, x_3 and y_1, y_2, y_3 and take the respective radii to be $(1-2^{1-p})^{1/p}$, $2^{(1-p)/p}$ and $2^{(1-p)/p}$. $\square$

§12. Extension of Contractions for Pairs of Banach Spaces

As we have just seen, it is always possible to extend contractions on Hilbert space, but not on ℓ_n^p when $n > 1$, $1 < p < \infty$ and $p \neq 2$. When X and Y are Banach spaces of dimension greater than 1, the question of whether the pair (X,Y) has the contraction extension property divides in much the same way provided Y is *strictly convex*, that is, $\|x+y\| = \|x\| + \|y\|$ $(y \neq 0)$ only if $x = \dfrac{\|x\|}{\|y\|} y$. The general solution is due to Schönbeck [84], but the simpler argument of Edelstein and Thompson [25] can be given in the special case $X = Y$. The argument employs a result in [24] concerning contractions on strictly convex spaces.

Lemma 12.1. *If X is a strictly convex Banach space, $T : X \to X$ is a contraction and $D \subseteq X$ is such that $T|_D$ is an isometry, then T is affine on co D (i.e.*

$$T\left(\sum_{i=1}^{n} \lambda_i x_i\right) = \sum_{i=1}^{n} \lambda_i T x_i \text{ if } x_i \in D, \lambda_i \geq 0 \ (1 \leq i \leq n) \text{ and } \sum_{i=1}^{n} \lambda_i = 1).$$

Proof. We will use induction on the number of elements in the convex combination. Suppose $x, y \in D$, $0 < \lambda < 1$ and $z = \lambda x + (1-\lambda)y$. Then $\|x - y\| = \|Tx - Ty\|$ and $\|Tz - Ty\| \leq \|z - y\|$, so that

$$\|x - z\| = (1-\lambda)\|x - y\| = \|x - y\| - \|z - y\| \leq \|Tx - Ty\| - \|Tz - Ty\|$$

$$\leq \|Tx - Tz\| \leq \|x - z\|.$$

Hence equality holds throughout and, in particular,

$$\|Tx - Tz\| = \|x - z\| \text{ and } \|Tz - Ty\| = \|z - y\|.$$

Thus

$$\|Tx - Tz\| + \|Tz - Ty\| = \|x - z\| + \|z - y\| = \|x - y\| = \|Tx - Ty\|$$

and, by virtue of the strict convexity,

$$Tx - Tz = \frac{\|Tx - Tz\|}{\|Tz - Ty\|}(Tz - Ty) = \frac{1-\lambda}{\lambda}(Tz - Ty).$$

Therefore the lemma holds if $n = 2$.

Suppose the result is true for convex combinations of n elements of D ($n \geq 2$) and let $x_1, \ldots, x_{n+1} \in D$ and $\sum_{i=1}^{n+1} \lambda_i = 1$, $\lambda_i > 0$ ($1 \leq i \leq n+1$). Let $\lambda = \sum_{i=1}^{n-1} \lambda_i$, $x = \sum_{i=1}^{n-1} \lambda_i x_i + (1-\lambda)x_n$ and $y = \sum_{i=1}^{n-1} \lambda_i x_i + (1-\lambda)x_{n+1}$. Since x and y are convex combinations of n elements of D, we have

$$\|Tx - Ty\| = \|(1-\lambda)Tx_n - (1-\lambda)Tx_{n+1}\| = (1-\lambda)\|x_n - x_{n+1}\| = \|x - y\|.$$

The lemma holds when $n = 2$, therefore

$$T\left(\frac{\lambda_n}{1-\lambda}x + \left(1 - \frac{\lambda_n}{1-\lambda}\right)y\right) = \frac{\lambda_n}{1-\lambda}Tx + \left(1 - \frac{\lambda_n}{1-\lambda}\right)Ty$$

$$= \sum_{i=1}^{n-1} \lambda_i Tx_i + \lambda_n Tx_n + \left(1 - \frac{\lambda_n}{1-\lambda}\right)(1-\lambda)Tx_{n+1}.$$

This is the desired result since $1 - \lambda - \lambda_n = \lambda_{n+1}$. $\square$

Theorem 12.2. *If X is a strictly convex Banach space, then (X,X) has the contraction extension property if and only if X is a Hilbert space.*

Proof. By Theorem 11.3, it suffices to show that if X is strictly convex and not a Hilbert space, then (X,X) does not have the contraction extension property. Suppose X is not a Hilbert space. Then, by a characterization of Hilbert space due to Lorch [55], there exist $x, y \in X$ such that $\|x\| = \|y\| = 1$ and $\|2x - y\| < \|2y - x\|$. Define $T : \{0, x, y\} \to X$ by $T(0) = 0$, $Tx = y$ and $Ty = x$. Clearly T is an isometry on $\{0, x, y\}$. If T has an extension $\tilde{T}$ to all of X as a contraction, then, by Lemma 12.1, we must have $\tilde{T}(y/2) = \frac{1}{2}\tilde{T}(y) = x/2$. But then $\|\tilde{T}(y/2) - \tilde{T}(x)\| = \|x/2 - y\| > \|x - y/2\|$, contradicting the assumption that $\tilde{T}$ is a contraction. Thus T cannot be extended to all of X and (X,X) does not have the contraction extension property. $\square$

The general result when $X \neq Y$ but Y is strictly convex requires more work. We state the theorem of Schönbeck [84] and then give a sequence of lemmas from which the theorem will follow.

Theorem 12.3. *Suppose X and Y are Banach spaces, Y is strictly convex and $\dim(Y) \geq 2$. Then (X,Y) has the contraction extension property if and only if X and Y are Hilbert spaces.*

Lemma 12.4. *If X and Y are Banach spaces, Y is strictly convex and (X,Y) has the contraction extension property, then the following condition holds:*

> *If $x_1, x_2 \in X$, $y_1, y_2 \in Y$, $\|x_1\| = \|y_1\|$, $\|x_2\| = \|y_2\|$ and $\|x_1 - x_2\| = \|y_1 - y_2\|$, then $\|ay_1 + by_2\| \leq \|ax_1 + bx_2\|$ for all real a, b.*   (12.1)

Proof. We first establish (12.1) for the case $a, b > 0$, $a + b = 1$. Define $T: \{0, x_1, x_2\} \to Y$ by $T(0) = 0$, $Tx_1 = y_1$ and $Tx_2 = y_2$. Obviously T is a contraction and by assumption T can be extended to $x = ax_1 + bx_2$. Thus there exists $y \in Y$ such that

$$\|y\| \leq \|x\| = \|ax_1 + bx_2\|,$$
$$\|y - y_1\| \leq \|x - x_1\| = b\|x_1 - x_2\| \text{ and}$$
$$\|y - y_2\| \leq \|x - x_2\| = a\|x_1 - x_2\|.$$

From $\|y_1 - y_2\| \leq \|y_1 - y\| + \|y - y_2\| \leq (a + b)\|x_1 - x_2\| = \|y_1 - y_2\|$ and the strict convexity of Y, it follows that

$$y_1 - y = \frac{\|y - y_1\|}{\|y - y_2\|}(y - y_2) = (b/a)(y - y_2), \text{ or}$$

$$y = ay_1 + by_2.$$

The other cases follow by considering multiplies of a and b and the obvious maps from $\{0, -x_1, -x_1 + x_2\}$ to $\{0, -y_1, -y_1 + y_2\}$ and from $\{0, -x_2, -x_2 + x_1\}$ to $\{0, -y_2, -y_2 + y_1\}$. □

Definition 12.5. *Let x and y be elements of a normed linear space. If $\|x + \lambda y\| \geq \|x\|$ for all real λ, x is said to be normal to y, or, in symbols, xNy.*

Lemma 12.6. *Let $\|\cdot\|$ be a norm on $Y = \mathbb{R} \times \mathbb{R}$.*

There exists $x, y \in Y$, $\|x\| = \|y\| = 1$, such that xNy and yNx. (12.2)

If $x \in Y$, there exists $y \in Y$ such that xNy and $y \neq 0$. (12.3)

There exists $x, y \in Y$, $\|x\| = \|y\| = 1$, such that $\|x + \lambda y\| > 1$ if $\lambda \neq 0$. (12.4)

Suppose $x, y \in Y$, xNy, $\|x\| = \|y\| = 1$ and $\{\gamma_n\}_{n=1}^{\infty}$ is a positive number sequence with limit zero. There exists sequences $\{x_n\}_{n=1}^{\infty}$ and $\{y_n\}_{n=1}^{\infty}$ in Y such that $\|x_n\| = \|y_n\| = 1$ $(n \geq 1)$, $\lim x_n = \lim y_n = x$ and $y_n - x_n = \gamma_n y$ for all large n. (12.5)

Proof. (12.2) Let C be the set of points of Y of norm 1 and define $f: C \times C \to \mathbb{R}$ by setting $f(x, y)$ equal to the area of the parallelogram determined by x and y. Let (x, y) be a pair in $C \times C$ at which f assumes its maximum. It is easy to see that the straight line through x (y) parallel to y (x) cannot intersect the interior of C. Hence xNy and yNx.

(12.3) Fix $x \in Y$ and let L be any line of support to the ball of radius $\|x\|$ at x. Then xNy for any y parallel to L.

(12.4) Let S be the Euclidean sphere of least radius which contains C and choose some $x \in C \cap S$. Then if $y \in C$ is parallel to the tangent line to S at x, clearly $\|x + \lambda y\| > 1$ for all $\lambda \neq 0$.

(12.5) For the function $g(\lambda) = \|x + \lambda y\|$, either (a) $g(\lambda)$ is strictly increasing for $\lambda \geq 0$, or (b) there exists $\lambda_0 > 0$ such that $g(\lambda) = 1$ for $0 \leq \lambda \leq \lambda_0$. If (a) does not hold then for some λ_1, λ_0 with $0 \leq \lambda_1 < \lambda_0$ we have $1 \leq \|x + \lambda_1 y\| = \|x + \lambda_0 y\| = r$. Choose α, $0 < \alpha \leq 1$, so that $\frac{1 - \alpha}{1 + (r - 1)\alpha} = \lambda_1/\lambda_0$. Then $\|\alpha r x + (1 - \alpha)(x + \lambda_0 y)\| \leq$

$\alpha r + (1-\alpha)r = r$, but $\|\alpha rx + (1-\alpha)(x+\lambda_0 y)\| = (1+(r-1)\alpha)\|x+\lambda_1 y\| = (1+(r-1)\alpha)r$. Hence $1+(r-1)\alpha \le 1$, or $r \le 1$. Thus $r=1$, $g(\lambda)=1$ for $0 \le \lambda \le \lambda_0$ and (b) holds.

In this situation, set $x_n = x$ and $y_n = \dfrac{x+\gamma_n y}{\|x+\gamma_n y\|}$. Clearly $\|x_n\| = \|y_n\| = 1 \ (n \ge 1)$, $\lim x_n = x$, $\lim y_n = x$, and if $\gamma_n \le \lambda_2$, $y_n - x_n = x + \gamma_n y - x = \gamma_n y$.

In case condition (a) holds, let $h(\lambda) = g^{-1}(\|x-\lambda y\|) \ (\lambda \ge 0)$. Then

$$u(\lambda) = \frac{\|h(\lambda)y + \lambda y\|}{\|x-\lambda y\|} = \frac{h(\lambda)+\lambda}{\|x-\lambda y\|}$$

is continuous, $u(0)=0$ and $\lim_{\lambda \to \infty} \inf u(\lambda) = 2$. Hence there exists a positive sequence $\{\lambda_n\}$ tending to 0 such that $u(\lambda_n) = \gamma_n$ for all large n. With the definitions

$$x_n = \frac{x-\lambda_n y}{\|x-\lambda_n y\|} \quad \text{and} \quad y_n = \frac{x+h(\lambda_n)y}{\|x-\lambda_n y\|} \quad (n \ge 1),$$

$\lim x_n = x$, $\lim y_n = x$, $\|x_n\| = 1$ and

$$\|y_n\| = \frac{\|x+h(\lambda_n)y\|}{\|x-\lambda_n y\|} = \frac{\|x+g^{-1}(\|x-\lambda_n y\|)y\|}{\|x-\lambda_n y\|} = 1 \ (n \ge 1).$$

Furthermore,

$$y_n - x_n = \frac{h(\lambda_n)y + \lambda_n y}{\|x-\lambda_n y\|} = u(\lambda_n)y = \gamma_n y \quad \text{for}$$

all large n. $\square$

Lemma 12.7. *Let X and Y be Banach spaces of dimension greater than 1 satisfying (12.1). If $x_1, x_2 \in X$ and $y_1, y_2 \in Y$ such that $\|x_1\| = \|x_2\| = \|y_1\| = \|y_2\| = 1$, $x_1 N x_2$ and $y_1 N y_2$, then $\|ay_1 + by_2\| \le \|ax_1 + bx_2\|$ for all real a, b. Moreover normality is symmetric in both X and Y.*

Proof. By (12.5) there exist sequences $\{x_n'\}_{n=1}^{+\infty}$ and $\{x_n''\}_{n=1}^{+\infty}$ in the linear span of x_1 and x_2 and sequences $\{y_n'\}_{n=1}^{+\infty}$ and $\{y_n''\}_{n=1}^{+\infty}$ in the linear span of y_1 and y_2 such that $\lim x_n' = \lim x_n'' = x_1$, $\lim y_n' = \lim y_n'' = y_1$, and $\|x_n'\| = \|x_n''\| = \|y_n'\| = \|y_n''\| = 1$, $x_n'' - x_n' = n^{-1}x_2$ and $y_n'' - y_n' = n^{-1}y_2$ for all large n. It follows from (12.1) that $\|ay_n' + by_n''\| \le \|ax_n' + bx_n''\|$ for $n \ge 1$, a, b real. From the above relations,

$$\|ax_1 + bx_2\| = \lim \|ax_n' + bx_2\| = \lim \|ax_n' + nb(x_n'' - x_n')\|$$
$$\ge \lim \|ay_n' + nb(y_n'' - y_n')\| = \lim \|ay_n' + by_2\| = \|ay_1 + by_2\|.$$

In showing normality is symmetric, it suffices to consider only unit vectors. Suppose $x_1, x_2 \in X$ with $\|x_1\| = \|x_2\| = 1$ and $x_1 N x_2$. By (12.2) there exist $y_1, y_2 \in Y$ of norm 1 such that $y_1 N y_2$ and $y_2 N y_1$. By the argument just completed we have, for all λ,

$$\|x_2 + \lambda x_1\| \ge \|y_2 + \lambda y_1\| \ge \|y_2\| = \|x_2\|.$$

Hence $x_2 N x_1$ and normality is symmetric in X.

To show the same is true in Y, choose $y_1, y_2 \in Y$, $\|y_2\| = \|y_1\| = 1$ and $y_1 N y_2$. According to (12.4), we may choose $x_1, x_2 \in X$, $\|x_1\| = \|x_2\| = 1$, such that

$\|x_1 + \lambda x_2\| > 1$ if $\lambda \neq 0$. By (12.3) there exists $y = ay_1 + by_2 \neq 0$ such that $y_2 N y$. Then

$$\|x_2 + \lambda(ax_1 + bx_2)\| = \|\lambda ax_1 + (1 + \lambda b)x_2\| \geq \|\lambda ay_1 + (1 + \lambda b)y_2\|$$
$$= \|y_2 + \lambda y\| \geq \|y_2\| = 1.$$

Therefore $x_2 N(ax_1 + bx_2)$ and, as normality is symmetric in X, we have

$$\|ax_1 + bx_2\| \leq \|ax_1 + bx_2 + (-b)x_2\| = |a|$$

But $\|ax_1 + bx_2\| = |a| \; \|x_1 + ba^{-1}x_2\| > |a|$ if $b \neq 0$. Hence $b = 0$ and $y_2 N y_1$. $\square$

Lemma 12.8. *Let $\|\cdot\|$ be a norm on $X = \mathbb{R} \times \mathbb{R}$ for which normality is symmetric, $\|(1,0)\| = \|(0,1)\| = 1$ and $(0,1) \, N(1,0)$. Then*

$$(a,b)N(c,d) \quad \text{if and only if} \quad |ad - bc| = \|(a,b)\| \; \|(c,d)\| \tag{12.6}$$

and

$$\|(a,b)\| = \|(b, -a)\|^* \quad \text{where} \quad \|\cdot\|^* \text{ is the usual norm in } X^*. \tag{12.7}$$

Proof. We begin by giving X a new norm, $\|\cdot\|_1$, defined by

$$\|(a,b)\|_1 = \begin{cases} \|(a,b)\| & \text{if } ab \geq 0 \\ \\ \|(b, -a)\|^* & \text{if } ab < 0. \end{cases}$$

That this is, in fact, a norm follows from the observation that the set $\{(a,b): \|(a,b)\|_1 \leq 1\}$ is convex and symmetric. Let X_1 denote X with norm $\|\cdot\|_1$.

We first show normality is symmetric in X_1. As usual we identify the dual X_1^* of X_1 with ordered number pairs $f = (p,q)$ where the action on X_1 is given by $f(a,b) = ap + bq$. The set $\{(c,d): f(c,d) = \|f\|_1^* \; \|(a,b)\|_1\}$ is a line of support to the sphere of radius $\|(a,b)\|_1$ at (a,b) if and only if

$$ap + bq = f(a,b) = \|(p,q)\|_1^* \; \|(a,b)\|_1. \tag{12.8}$$

Therefore $(a,b) \, N(c,d)$ in X_1 if and only if $pc + qd = 0$ for some (p,q) satisfying (12.8). But $pc + qd = 0$ implies (p,q) is a multiple of $(d, -c)$; hence $(a,b) \, N(c,d)$ if and only if $|ad - bc| = \|(d, -c)\|_1^* \; \|(a,b)\|_1$. If $cd \leq 0$,

$$\|(d, -c)\|_1^* = \sup_{rs > 0} \frac{|rd - sc|}{\|(r,s)\|_1} = \sup_{rs > 0} \frac{|rd - sc|}{\|(r,s)\|} = \|(d, -c)\|^* = \|(c,d)\|_1,$$

and if $cd \geq 0$,

$$\|(d, -c)\|_1^* = \sup_{rs > 0} \frac{|rd - sc|}{\|(r,s)\|_1} = \sup_{rs > 0} \frac{|rd + sc|}{\|(r, -s)\|_1} = \sup_{rs > 0} \frac{|rd + sc|}{\|(-s, -r)\|^*}$$
$$= \|(c,d)\| = \|(c,d)\|_1.$$

Thus $(a,b)N(c,d)$ if and only if $|ad - bc| = \|(a,b)\|_1 \; \|(c,d)\|_1$. Hence normality is symmetric in X_1, and $\|\cdot\|_1$ is a norm on X_1 which satisfies (12.6) and (12.7).

To complete the proof we show that $\|\cdot\| = \|\cdot\|_1$. We need only show that $\|(a,b)\| = \|(a,b)\|_1$ when $ab < 0$. Suppose $r = f(\theta)$ and $r = \varphi(\theta)$ are the equations of the unit circles in $\|\cdot\|_1$ and $\|\cdot\|$ respectively, where $0 \leq \theta \leq 2\pi$. These functions are Lipschitzian, for if $0 \le \theta_1 < \theta_2 \le \pi/4$, then one of the following situations must occur:

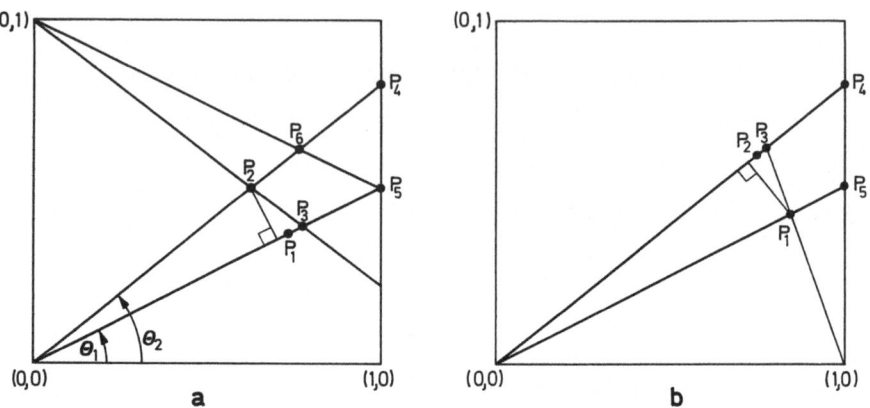

In each figure $P_1 = (\theta_1, r_1)$ and $P_2 = (\theta_2, r_2)$. Figure (a) represents the case $r_2 \le r_1$ and figure (b) represents $r_2 \ge r_1$. In either case, $|r_2 - r_1| \le |\overline{P_1 P_2}| \le K |\overline{P_4 P_5}|$, where K is an absolute constant. This is obvious in figure (b) and in (a) we have $|\overline{P_2 P_3}| \le |\overline{P_5 P_6}|$ and since in $\triangle P_4 P_5 P_6$ all angles are between $\pi/4$ and $\pi/2$, we obtain the constant K. But then $|\overline{P_4 P_5}| = |\tan \theta_2 - \tan \theta_1| \sec^2 \xi$ where $0 \le \xi \le \pi/4$. Lipschitzian functions are absolutely continuous, hence the cotangent of the angle between the vector $(\theta, f(\theta))$ (polar coordinates) and the tangent vector at $(\theta, f(\theta))$ is $f'(\theta)/f(\theta)$ whenever $f'(\theta)$ exists. A similar statement holds for $\varphi(\theta)$. That these functions are Lipschitzian also follows from Klee [45].

Suppose $\pi/2 \le \theta \le \pi$. By (12.3) there exists τ, $0 \le \tau \le \pi/2$, such that $(\theta, f(\theta)) N(\tau, f(\tau))$ and $(\tau, f(\tau)) N(\theta, f(\theta))$, since normality is symmetric. Because the two circles agree in quadrant one, we have $(\tau, \varphi(\tau)) N(\theta, \varphi(\theta))$ and again by symmetry, $(\theta, \varphi(\theta)) N(\tau, \varphi(\tau))$. Thus $f'(\theta)/f(\theta) = \varphi'(\theta)/\varphi(\theta)$ for almost all $\theta \in [0, \pi]$. Since $\log f/\varphi$ is absolutely continuous and $\dfrac{d}{d\theta}\left(\log f/\varphi\right)(\theta) = f'(\theta)/f(\theta) - \varphi'(\theta)/\varphi(\theta) = 0$ almost everywhere, it follows that f/φ is constant. Now $f = \varphi$ on $[0, \pi/2]$, so $f = \varphi$ on $[0, \pi]$ and by symmetry, $\|\cdot\| = \|\cdot\|_1$. $\square$

Proof of Theorem 12.3. Choose points $x_1, x_2 \in X$ and $y_1, y_2 \in Y$ all of norm 1 such that $x_1 N x_2$ and $y_1 N y_2$. Identify x_1 and y_1 with $(1,0)$ and x_2 and y_2 with $(0,1)$ and let $\|\cdot\|_1$ and $\|\cdot\|_2$ be the norms on $B = \mathbb{R} \times \mathbb{R}$ induced by the norms in X and Y respectively. According to Lemma 12.7, normality is symmetric in both norms and

$$\|(a,b)\|_2 \le \|(a,b)\|_1 \text{ for all real } a, b. \tag{12.9}$$

Moreover, by lemma 12.8,

$$\|(a,b)\|_2 = \sup_{(c,d) \ne 0} \frac{|ac + bd|}{\|(c,d)\|_2^*} = \sup_{(c,d) \ne 0} \frac{|ac + bd|}{\|(d,-c)\|_2}$$

$$\geq \sup_{(c,d)\neq 0} \frac{|ac+bd|}{\|(d,-c)\|_1} = \sup_{(c,d)\neq 0} \frac{|ac+bd|}{\|(c,d)\|_1^*} = \|(a,b)\|_1$$

for all real a and b and therefore $\|\cdot\|_1 = \|\cdot\|_2$. From equality of norms and by interchanging y_1 and y_2 and also y_2 and $-y_2$, we obtain

$$\|(a,b)\|_1 = \|(b,a)\|_1 = \|(a,-b)\|_1 \text{ for all real } a, b. \tag{12.10}$$

Suppose $\|(a,b)\|_1 = \|(c,d)\|_1$ and γ,β are real. From Lemma 12.4 it follows that

$$\|\gamma(a,b) + \beta(c,d)\|_1 \geq \|\gamma(c,d) + \beta(a,b)\|_2 = \|\gamma(c,d) + \beta(a,b)\|_1,$$

and hence, by symmetry,

$$\|\gamma(a,b) + \beta(c,d)\|_1 = \|\gamma(c,d) + \beta(a,b)\|_1.$$

Furthermore if

$$\|(a,b)-(c,d)\|_1 = \|(a,b)+(c,d)\|_1,$$

then

$$\|(\gamma+\beta)(a,b)+(\gamma-\beta)(c,d)\|_1 = \|\gamma((a,b)+(c,d))+\beta((a,b)-(c,d))\|_1$$
$$= \|\beta((a,b)+(c,d))+\gamma((a,b)-(c,d))\|_1 = \|(\gamma+\beta)(a,b)-(\gamma-\beta)(c,d)\|_1.$$

The result of dividing this last expression by $\gamma+\beta$ is

$$\|(a,b) + \lambda(c,d)\|_1 = \|(a,b) - \lambda(c,d)\|_1 \text{ for all real } \lambda. \tag{12.11}$$

By (12.10),

$$\|(a,b)\|_1 = \|(b,-a)\|_1$$

and

$$\|(a,b) + (b,-a)\|_1 = \|(a+b,b-a)\|_1 = \|(a-b,a+b)\|_1 = \|(a,b)-(b,-a)\|_1.$$

Thus, by (12.11), (a,b) $N(b,-a)$. Applying (12.6) of Lemma 12.8, we obtain

$$a^2+b^2 = \|(a,b)\|_1 \|(b,-a)\|_1 = \|(a,b)\|_1^2.$$

Hence all two dimensional subspaces of X and Y are Euclidean, equivalent to the assertion X and Y are Hilbert spaces. $\square$

The contraction extension problem for pairs of Banach spaces (X,Y) appears to be difficult when Y is not strictly convex. However, in case $X = Y$ there is a partial solution. According to Theorem 11.2 the pair (X,X) has the contraction extension property if X has the binary intersection property. In fact, these are the only such pairs having the contraction extension property among the separable dual spaces which are not strictly convex. As a prelude to formally stating this result we give two theorems dealing with the binary intersection property, due to Hanner [35] in finite dimensions and Lindenstrauss [53] in a general Banach space. We need only the following restricted version.

Theorem 12.9. *Let X be a real dual Banach space and choose an integer $n > 2$. If every collection of n mutually intersecting balls of equal radius has nonempty intersection, then every collection of n mutually intersecting balls has nonempty intersection.*

Proof. Suppose the conclusion is false. Let $k \leq n$ be the smallest positive integer such that there exists a mutually intersecting collection $\{B(x_i, r_i)\}_{i=1}^{k}$ of balls in X with $\bigcap_{i=1}^{k} B(x_i, r_i) = \emptyset$. By virtue of weak* compactness we must have $\bigcap_{i=1}^{k} B(x_i, r_i + \epsilon) = \emptyset$ for some $\epsilon > 0$. Put $r = 1 + \max_{1 \leq i \leq k} r_i$. We inductively establish the existence of balls $B(y_i, r) \supset B(x_i, r_i)$ $(1 \leq i \leq k)$ such that $\bigcap_{i=1}^{k} B(y_i, r) = \emptyset$, providing the desired contradiction.

Suppose that we have determined y_i for $i \leq j \leq k$ (j possibly 0) such that $B(y_i, r) \supset B(x_i, r_i)$ and $(\bigcap_{i=1}^{j} B(y_i, r)) \cap (\bigcap_{i=j+1}^{k} B(x_i, r_i + \epsilon)) = \emptyset$. (We assume the empty intersection is X.) Let

$$K = (\bigcap_{i=1}^{j} B(y_i, r)) \cap (\bigcap_{i=j+2}^{k} B(x_i, r_i + \epsilon)).$$

Then K is a compact convex set disjoint from the compact convex set $B(x_{j+1}, r_{j+1} + \epsilon)$. Thus there exists $f \in X^*$, $\|f\| = 1$, separating K and $B(x_{j+1}, r_{j+1} + \epsilon)$. Now $f(x - x_{j+1})$ takes all values in $(-r_{j+1} - \epsilon, \ r_{j+1} + \epsilon)$ for $x \in B(x_{j+1}, r_{j+1} + \epsilon)$ and therefore $f(x - x_{j+1}) > r_{j+1} + \epsilon$ for $x \in K$ (possibly replace f with $-f$). Let $z \in X$ such that $\|z\| = 1$ and $f(z) \leq -1 + \epsilon/(r - r_{j+1})$. Set $y_{j+1} = x_{j+1} + (r - r_{j+1})z$. If $x \in B(x_{j+1}, r_{j+1})$, then

$$\|x - y_{j+1}\| \leq \|x - x_{j+1}\| + \|x_{j+1} - y_{j+1}\| \leq r_{j+1} + (r - r_{j+1}) = r,$$

and therefore $B(x_{j+1}, r_{j+1}) \subset B(y_{j+1}, r)$. However if $x \in B(y_{j+1}, r)$, then $r \geq f(x - y_{j+1}) = f(x - x_{j+1}) - (r - r_{j+1})f(z)$. Hence $f(x - x_{j+1}) \leq r + (r - r_{j+1})(-1 + \epsilon/(r - r_{j+1})) = r_{j+1} + \epsilon$, so that $x \notin K$ and $B(y_{j+1}, r) \cap K = \emptyset$. $\square$

Theorem 12.10. *Let X be a Banach space such that every collection of four mutually intersecting balls in X has nonempty intersection. Then every finite collection of mutually intersecting balls has nonempty intersection.*

Proof. We will induct on the number n of balls. Suppose the result is true for n $(n \geq 4)$ and $\{B_i\}_{i=1}^{n+1}$ is a collection of mutually intersecting balls in X. Let $x_1 \in X$, $\theta = \max_{1 \leq i \leq n+1} d(x_1, B_i)$ and $B_0 = B(x_1, \theta + \epsilon)$, where $\epsilon > 0$ will be chosen later. The collection $\{B_i\}_{i=0}^{n+1}$ is also mutually intersecting. Let $A = \{1, 2, \ldots, n+1\}$ and denote by Ω the set of all subsets of A consisting of $n - 1$ elements. The cardinality of Ω is $n(n+1)/2$. By assumption, for $\alpha \in \Omega$, there exists $y_\alpha \in B_0 \cap (\bigcap_{i \in \alpha} B_i)$. Let

$$y = \frac{2}{n(n+1)} \sum_{\alpha \in \Omega} y_\alpha.$$

Then $y_\alpha \in B_0$ for each α and B_0 is convex, so $y \in B_0$ and $\|y - x_1\| \leq \theta + \epsilon$. We estimate $d(y, B_i)$ for $1 \leq i \leq n+1$. The number of $\alpha \in \Omega$ for which $i \in \alpha$ is $n(n-1)/2$. Therefore

$$d(y, B_i) \leq \frac{2}{n(n+1)} \sum_{\alpha \in \Omega} d(y_\alpha, B_i) = \frac{2}{n(n+1)} \sum_{i \notin \alpha} d(y_\alpha, B_i)$$

$$\leq \frac{2}{n(n+1)} \sum_{i \notin \alpha} (d(x_1, B_i) + \|y_\alpha - x_1\|)$$

$$\leq \frac{2}{n(n+1)} \left(\frac{n(n+1)}{2} - \frac{n(n-1)}{2} \right) (\theta + \theta + \varepsilon)$$

$$= \frac{4}{n+1} (\theta + \varepsilon/2).$$

Choose C so that $\frac{4}{n+1} < C < 1$. For a suitable choice of ε we have $d(y,B_i) < C\theta$ $(1 \leq i \leq n+1)$ and $\|y - x_1\| < 2\theta$. Letting $x_2 = y$ and repeating the argument, we find $x_3 \in X$ such that $d(x_3,B_i) \leq C^2\theta$ $(1 \leq i \leq n+1)$ and $\|x_2 - x_3\| < 2C\theta$. In general there exists $\{x_m\}_{m=1}^{\infty} \subseteq X$ such that $d(x_m,B_i) \leq C^{m-1}\theta$ $(1 \leq i \leq n)$ and $\|x_m - x_{m+1}\| < 2C^{m-1}\theta$. The sequence $\{x_m\}_{m=1}^{\infty}$ is Cauchy and its limit x satisfies $d(x,B_i) = 0$ $(1 \leq i \leq n+1)$. $\square$

Theorem 12.11. *Suppose X is a separable dual Banach space. Then (X,X) has the contraction extension property if and only if X is a Hilbert space or X has the binary intersection property.*

Theorem 12.11 is due to B. Grünbaum [32] if dim $X = 2$ and to Schönbeck [84] if X is finite-dimensional. The proof will be broken into a sequence of lemmas.

Lemma 12.12. *Suppose X is a Banach space such that (X,X) has the contraction extension property. If e_1 and e_2 are extreme points of the unit ball B of X and $x_1, x_2 \in X$ are such that $\|x_1 + e_1\| = \|x_2 + e_2\|$ and $\|x_1 - e_1\| = \|x_2 - e_2\|$ then*

$$\|\lambda x_1 + \mu e_1\| = \|\lambda x_2 + \mu e_2\| \quad for \quad |\lambda| \geq |\mu|.$$

Proof. It suffices to prove $\|x_1 + \mu e_1\| = \|x_2 + \mu e_2\|$ for $0 \leq \mu \leq 1$. Define T by $T(x_1 - e_1) = x_2 - e_2$, $T(x_1 + e_1) = x_2 + e_2$ and $T(0) = 0$. Then T is a contraction and can be extended to $x_1 + \mu e_1$. Let $y = T(x_1 + \mu e_1)$. We have

$$\|y - (x_2 - e_2)\| \leq \|x_1 + \mu e_1 - (x_1 - e_1)\| = 1 + \mu \qquad \text{and} \qquad (12.12)$$

$$\|y - (x_2 + e_2)\| \leq \|x_1 + \mu e_1 - (x_1 + e_1)\| = 1 - \mu. \qquad (12.13)$$

Since $\|x_2 - e_2 - (x_2 + e_2)\| = 2$, we must have equality in (12.12) and (12.13) and then, since e_2 is an extreme point, it follows that $y = x_2 + \mu e_2$. Thus $\|x_2 + \mu e_2\| \leq \|x_1 + \mu e_1\|$ and the lemma follows by symmetry. $\square$

Lemma 12.13. *Suppose X is a Banach space and (X,X) has the contraction extension property. If Y is a two dimensional subspace of X containing an extreme point e of the unit ball of X, there exists $y \in Y$ such that $\|y\| = 1$ and $\|\lambda y + \mu e\| = \|\lambda y - \mu e\|$ for all real λ, μ. Furthermore if Y' is any other two dimensional subspace with y' and e' as above, then the linear map $T : Y \to Y'$ defined by $T(e) = e'$ and $T(y) = y'$ is an isometry.*

Proof. For $n \geq 1$ there exists $y_n \in Y$ such that $\|y_n\| = 1$ and $\|y_n - ne\| = \|y_n + ne\|$. Applying Lemma 12.12 with $e_1 = e$, $e_2 = -e$ and $x_1 = x_2 = \frac{y_n}{n}$, we obtain

$$\left\| \frac{\lambda}{n} y_n + \mu e \right\| = \left\| \frac{\lambda}{n} y_n - \mu e \right\| \text{ for } |\lambda| \geq |\mu|, \text{ or}$$

$$\|\lambda y_n + \mu e\| = \|\lambda y_n - \mu e\| \quad \text{for } |\lambda| \geq \frac{|\mu|}{n}.$$

Clearly any limit point of $\{y_n\}_{n=1}^{\infty}$ satisfies the required conditions.

Now suppose y, y', e and e' are as in the lemma. Note that

$$1 = \|e\| \leq \tfrac{1}{2}(\|\lambda y + e\| + \|\lambda y - e\|) = \|\lambda y + e\| \qquad (\lambda \in \mathbb{R}).$$

For $n \geq 1$, there exists $\varepsilon_n \geq 0$ such that $\|\frac{y}{n} + e\| = \|\varepsilon_n y' + e'\|$ and therefore $\|\frac{y}{n} - e\| = \|\varepsilon_n y' - e'\|$. By Lemma 12.12 we have

$$\left\|\lambda \frac{y}{n} - \mu e\right\| = \|\lambda \varepsilon_n y' - \mu e'\| \quad \text{for } |\lambda| \geq |\mu|. \tag{12.14}$$

With $\mu = 0$ and $\lambda = 1$ we see that $\varepsilon_n = 1/n$ and the lemma then follows by letting $n \to \infty$ (12.14). $\square$

Lemma 12.14. *Suppose X is a Banach space which is not strictly convex and (X,X) has the contraction extension property. There are positive numbers a and c with the following properties:*

> *if e_1 and e_2 are extreme points of the unit ball B of X, then* (12.15)
> *$\|e_1 - e_2\| \geq c$; and*

> *if Y is a two dimensional subspace of X containing an extreme* (12.16)
> *point e of B, then e is the endpoint of two line segments of length*
> *a on the surface of the unit ball of Y.*

Proof. Let $x_1, x_2 \in X$ such that $x_1 \neq x_2$, $\|x_1\| = \|x_2\| = 1$ and $\|x_1 + x_2\| = 2$. If e is an extreme point of B, choose $x \in X$ so that $\|e - x\| = \|x_1 - x_2\|$ and $\|e + x\| = 2$. The map T defined by $T(x_1) = e$, $T(-x_1) = -e$ and $T(x_2) = x$ is a contraction and hence can be extended to 0. Since e is an extreme point of B, we must have $T(0) = 0$. Then $\|x\| = \|T(0) - T(x_2)\| \leq \|x_2\| \leq 1$ and $\|e + x\| = 2$ imply $\|x\| = 1$. Therefore $\|\lambda x + \mu e\| = \lambda + \mu$ for $\lambda, \mu \geq 0$.

Let Y be the two dimensional subspace of X determined by x and e, and choose $y \in Y$ such that $\|\lambda y + \mu e\| = \|\lambda y - \mu e\|$ for all λ and μ as guaranteed by Lemma 12.13. Then $x = \alpha e + \beta y$ for some $\alpha, \beta \in \mathbb{R}$. If $x' = \alpha e - \beta y$, then $\|x'\| = 1$ and $\|\lambda e + \mu x'\| = \|(\lambda + \mu\alpha)e - \mu\beta y\| = \|(\lambda + \mu\alpha)e + \mu\beta y\| = \|\lambda e + \mu x\| = \lambda + \mu \; (\lambda, \mu \geq 0)$. Thus the segments $[e, x]$ and $[e, x']$ belong to the boundary of the unit ball B_Y of Y. Let Γ denote the complementary arc to the two maximal segments in the boundary of B_Y containing e. Then $d(e, \Gamma) = c > 0$.

Suppose e_1 and e_2 are extreme points of B and $e_1 \neq \pm e_2$. Let Y' be the two dimensional subspace determined by e_1 and e_2. By Lemma 12.13, there is a linear isometry of Y' onto Y taking e_1 to e. Since e_2 is an extreme point of B, its image must lie in $\bar{\Gamma}$ and therefore $\|e_1 - e_2\| \geq c$, completing (12.15).

For (12.16) suppose e_1 and e_2 are extreme points of B and $x_1, x_2 \in B$ are such that $[e_1, x_1]$ and $[e_2, x_2]$ are maximal segments in the boundary of B (maximal with respect to having e_i as endpoint and direction determined by x_i). Let Y_1 and Y_2 be the two dimensional subspaces determined by e_1, x_1 and e_2, x_2 respectively. By Lemma 12.13, there exists $y_i \in Y_i$ such that $\|\lambda y_i + \mu e_i\| = \|\lambda y_i - \mu e_i\|$ for all

λ, μ and such that y_i belongs to the same half-plane of Y_i as x_i determined by the line through 0 and e_i $(i=1,2)$. Let T be the linear isometry with $T(e_1) = e_2$ and $T(y_1) = y_2$. Then T carries $[e_1, x_1]$ into a segment of the unit ball of Y_2 parallel to $[e_2, x_2]$ and since both segments are maximal, we must have $T(x_1) = x_2$. Therefore $\|e_1 - x_1\| = \|e_2 - x_2\|$. $\square$

Proof of Theorem 12.11. Suppose (X, X) has the contraction extension property and X is not strictly convex. By Theorems 11.2 and 11.3 it suffices to show that X has the binary intersection property. By Theorems 11.1, 12.9 and 12.10 it suffices to find balls $\{B(x_i, 1)\}_{i=1}^{4}$ in X such that

$$\|x_i - x_j\| = 2 \text{ if } i \neq j \text{ and } \bigcap_{i=1}^{4} B(x_i, 1) \neq \emptyset. \tag{12.17}$$

The case when X is two-dimensional will be handled separately. First, by Lemma 12.14, the boundary of the unit ball B of X is a polygon with sides of length a. Suppose z_1, z_2 and z_3 are consecutive vertices. There exists α, $0 \leq \alpha \leq 1$, such that $z_1 - \alpha z_2$ and $z_3 - \alpha z_2$ lie on a straight line through 0 and $\|z_3 - \alpha z_2\| \geq \|z_2 - \alpha z_2\|$. Thus $\|z_1 - \alpha z_2 - (z_2 - \alpha z_2)\| \leq \|z_1 - \alpha z_2\| + \|z_3 - \alpha z_2\| = \|z_1 - \alpha z_2 - (z_3 - \alpha z_2)\|$, or $\|z_1 - z_2\| \leq \|z_1 - z_3\|$. If $\|z_1 - z_2\| < \|z_1 - z_3\|$, then there exists v_1 and v_2 on the line through z_1 and z_2 symmetric with respect to $\frac{z_1 + z_2}{2}$ such that

$\|z_2 - z_1\| < \|v_2 - v_1\| \leq \|z_1 - z_3\|$ and such that $\|v_i + \frac{z_1 + z_2}{2}\| \leq 2$ $(i = 1,2)$. The

map T defined by $T(z_1) = v_1$, $T(z_3) = v_2$ and $T(-z_2) = -\frac{z_1 + z_2}{2}$ is a contraction

which cannot be extended to 0 since no ball of radius 1 contains v_1, v_2 and $-\frac{z_1 + z_2}{2}$.

Therefore $\|z_1 - z_2\| = \|z_1 - z_3\|$ and we will show this common value is 2.

Let w be the intersection of the line through 0 and z_2 with the line through z_1 and z_3. Assume, without loss of generality, that $\|z_1 - w\| \leq \frac{1}{2}\|z_1 - z_3\|$. Then $1 = \|z_1\| \leq \|z_1 - w\| + \|w\| \leq \frac{1}{2}\|z_1 - z_3\| + \|w\|$ and $\|z_1 - z_3\| = \|z_1 - z_2\| \leq \|z_1 - w\| + \|w - z_2\| \leq \frac{1}{2}\|z_1 - z_3\| + 1 - \|w\|$. Thus $1 \leq \frac{1}{2}\|z_2 - z_3\| + \|w\| \leq 1$ and $\|z_1 - w\| = \frac{1}{2}\|z_1 - z_3\|$. It follows that $w = \frac{z_1 + z_3}{2}$ and

$$1 = \left\|\frac{z_1 - z_3}{2} + \frac{z_1 + z_3}{2}\right\| = \left\|\frac{z_1 - z_3}{2}\right\| + \left\|\frac{z_1 + z_3}{2}\right\|.$$

Since z_1 is extreme and $z_1 = \frac{z_1 - z_3}{2} + \frac{z_1 + z_3}{2}$, we must have $\frac{z_1 + z_3}{2} = 0$.

Thus $\|z_1 - z_3\| = 2$ and the balls with centers z_1, $-z_1$, z_2 and $-z_2$ satisfy (12.17).

Now suppose dim $X > 2$ and let E be the set of extreme points of the unit ball B of X. Since X is separable and B is weak* compact, it follows from Choquet's Theorem [71] that each point of B can be represented by a positive Borel probability measure supported on E. Also, by separability and (12.15), we see that E is countable and therefore

$$B = \left\{ \sum_{n=1}^{\infty} \lambda_n e_n : \sum_{n=1}^{\infty} \lambda_n = 1,\ \lambda_n \geq 0 \text{ and } e_n \in E \ (n \geq 1) \right\}. \quad (12.18)$$

We use (12.18) to show that some plane face of the boundary of B contains three extreme points.

Let $e \in E$ and suppose $[e,x]$ is a maximal segment in the boundary of B. We may choose x so that $x \notin E$, for if $x \in E$ consider segments with direction near x and use (12.15). By (12.18), $x = \sum_{n=2}^{\infty} \lambda_n e_n$ where λ_2 and λ_3 are nonzero, $e_n \neq e$ $(n \geq 2)$, $e_2 \neq e_3$ and $\sum_{n=2}^{\infty} \lambda_n = 1$. Then, letting $e_1 = e$, $\beta_1 = 1/2$, and $\beta_n = \lambda_n/2 \ (n \geq 2)$, we have $1 = \left\| \frac{e+x}{2} \right\| = \left\| \sum_{n=1}^{\infty} \beta_n e_n \right\|$. Obviously $\left\| \frac{\beta_1 e_1 + \beta_2 e_2 + \beta_3 e_3}{\beta_1 + \beta_2 + \beta_3} \right\| = 1$ and therefore the triangle determined by e_1, e_2 and e_3 lies in the boundary of B. Let P be the plane determined by e_1, e_2 and e_3. We consider separately the cases $P \cap B = \Delta e_1 e_2 e_3$ and $P \cap B \supsetneq \Delta e_1 e_2 e_3$.

Suppose $x \in P \cap B$ and $x \notin \Delta e_1 e_2 e_3$. Assume, without loss of generality, that $[e_1,x]$ intersects $[e_2,e_3]$ and choose y so that $x \in [e_1,y]$ and $[e_1,y]$ is maximal in the boundary of B. If z is such that $y \in [e_2,z]$ and $[e_2,z]$ is maximal, then $[e_1,z]$, $[e_2,z]$ and $[e_3,z]$ are all maximal segments and $[e_1,z] \cap [e_2,e_3] \neq \emptyset$. Therefore,

$$\|e_1 - e_2\| = \|e_2 - e_3\| = \|e_1 - z\| = \|e_2 - z\| = \|e_3 - z\| = a. \quad (12.19)$$

Let w be the intersection of $[e_1,z]$ and $[e_2,e_3]$. Then $\|w - e_1\| = \|w - z\| = \|w - e_2\| = \|w - e_3\| = a/2$ or we contradict one of the equalities in (12.19). For example, if $\|w - z\| < a/2$, then either $\|w - e_2\| \leq a/2$ or $\|w - e_3\| \leq a/2$, in which case $\|z - e_i\| \leq \|z - w\| + \|w - e_i\| < a$ $(i = 2$ or $i = 3)$. Then the balls with centers $2e_1/a$, $2e_2/a$, $2e_3/a$, $2z/a$ and radius 1 have $2w/a$ in their intersection and therefore satisfy (12.17).

If $P \cap B = \Delta e_1 e_2 e_3$, then $\|e_1 - (\lambda e_2 + (1-\lambda)e_3)\| = a$ for $0 \leq \lambda \leq 1$, since the segment $[e_1, \lambda e_2 + (1-\lambda)e_3]$ is maximal. A similar statement holds for permutations of e_1, e_2 and e_3. The balls $B(e_1,a)$, $B(e_2 + e_3 - e_1, a)$ and $B(2e_2 - e_3, a)$ have distance between centers $2a$ and common point e_2. Thus, by Theorem 11.1, every collection of three mutually intersecting balls in X of equal radii has nonempty intersection. Let $y \in X$ such that $\|y\| = 1$ and $\|\alpha e_1 + \beta y\| = \|\alpha e_1 - \beta y\|$ $(\alpha, \beta \in R)$ as guaranteed by Lemma 12.13. Let $\lambda \geq 1$ so that $\|\lambda y + e_1\| = 2$. The balls $B(e_1,1)$, $B(-e_1,1)$ and $B(\lambda y,1)$ are mutually intersecting, hence $0 \in B(\lambda y,1)$. Thus $\lambda = 1$ and the balls $B(e_1,1)$, $B(-e_1,1)$, $B(y,1)$ and $B(-y,1)$ satisfy (12.17). $\square$

One should observe that the proofs of Theorems 12.2 and 12.3 only required the extension of contractions from 3-point sets to a fourth point and the proof of Theorem 12.11 only required extension from 4-point sets. In §13 we consider the extension problem with special restrictions on the range and domain.

§13. Special Extension Problems

We begin this section by considering contractions and isometries defined on convex sets. Edelstein and Thompson use Lemma 12.1 to obtain the following result for isometries on strictly convex spaces [25].

Theorem 13.1. *Let C be a convex subset of a strictly convex normed linear space X such that*

$$X = \left\{ \sum_{i=1}^{n} \lambda_i x_i : n \geq 1,\ x_i \in C,\ \lambda_i \in R \ \text{ and } \ \sum_{i=1}^{n} \lambda_i = 1 \right\}.$$

Then every isometry from C into X can be extended as an isometry to all of X.

Proof. Let $T : C \to X$ be an isometry. By Lemma 12.1, T is affine on C. Suppose $z \in X$ and write $z = \sum_{i=1}^{j} \lambda_i x_i + \sum_{i=1}^{k} \beta_i y_i$ where $\lambda_i > 0$ ($1 \leq i \leq j$), $\beta_i < 0$ ($1 \leq i \leq k$), $\sum_{i=1}^{j} \lambda_i + \sum_{i=1}^{k} \beta_i = 1$ and $x_i, y_i \in C$. Then $z = \lambda x + (1 - \lambda) y$ where $\lambda = \sum_{i=1}^{j} \lambda_i$, $x = \frac{1}{\lambda} \sum_{i=1}^{j} \lambda_i x \in C$ and $y = \frac{1}{1 - \lambda} \sum_{i=1}^{k} \beta_i y_i \in C$. Define $\tilde{T}$ by setting $\tilde{T}(z) = \lambda T(x) + (1 - \lambda) T(y)$. Observe that $\tilde{T}$ is uniquely defined, for if $z = \lambda' x' + (1 - \lambda') y'$ with $\lambda' \geq 1$, $x', y' \in C$, then $\lambda x - (1 - \lambda') y' = \lambda' x' - (1 - \lambda) y$. Hence if $\alpha = \lambda - (1 - \lambda') = \lambda' - (1 - \lambda) = \lambda' + \lambda - 1 > 0$, we have $\frac{\lambda}{\alpha} x + \frac{\lambda' - 1}{\alpha} y' = \frac{\lambda'}{\alpha} x' + \frac{\lambda - 1}{\alpha} y.$

Both sides are convex combinations of elements of C, so $\frac{\lambda}{\alpha} Tx + \frac{\lambda' - 1}{\alpha} Ty' = \frac{\lambda'}{\alpha} Tx' + \frac{\lambda - 1}{\alpha} Ty$ and the uniqueness of T follows. Clearly $\tilde{T}|_C = T$.

To see that $\tilde{T}$ is an isometry, let $z, w \in X$ and write $z = \lambda_1 x_1 + \lambda_2 x_2$, $w = \beta_1 y_1 + \beta_2 y_2$ where $\lambda_1, \beta_1 \geq 1$, $\lambda_1 + \lambda_2 = 1 = \beta_1 + \beta_2$, $x_1, x_2 \in C$ and $y_1, y_2 \in C$. Then

$$\|\tilde{T}z - \tilde{T}w\| = \|\lambda_1 Tx_1 + \lambda_2 Tx_2 - \beta_1 Ty_1 - \beta_2 Ty_2\| \tag{13.1}$$

$$= (\lambda_1 - \beta_2) \left\| \left(\frac{\lambda_1}{\lambda_1 - \beta_2} Tx_1 + \frac{-\beta_2}{\lambda_1 - \beta_2} Ty_2 \right) - \left(\frac{-\lambda_2}{\lambda_1 - \beta_2} Tx_2 + \frac{\beta_1}{\lambda_1 - \beta_2} Ty_1 \right) \right\|$$

Now $1 = \frac{\lambda_1}{\lambda_1 - \beta_2} + \frac{-\beta_2}{\lambda_1 - \beta_2} = \frac{-\lambda_2}{\lambda_1 - \beta_2} + \frac{\beta_1}{\lambda_1 - \beta_2}$, so (13.1) becomes

$$\|\tilde{T}z - \tilde{T}w\| = (\lambda_1 - \beta_2) \left\| T \left(\frac{\lambda_1}{\lambda_1 - \beta_2} x_1 + \frac{-\beta_2}{\lambda_1 - \beta_2} y_2 \right) - T \left(\frac{-\lambda_2}{\lambda_1 - \beta_2} x_2 + \frac{\beta_1}{\lambda_1 - \beta_2} y_1 \right) \right\|$$

$$= (\lambda_1 - \beta_2) \left\| \left(\frac{\lambda_1}{\lambda_1 - \beta_2} x_1 + \frac{-\beta_2}{\lambda_1 - \beta_2} y_2 \right) - \left(\frac{-\lambda_2}{\lambda_1 - \beta_2} x_2 + \frac{\beta_1}{\lambda_1 - \beta_2} y_1 \right) \right\|$$

$$= \|z - w\|. \quad \square$$

De Figueiredo and Karlovitz [28], [29] also considered contractions on convex sets but imposed further restrictions on the range of the extension. Precisely, they considered spaces X and closed convex subsets C with the following property:

if $T : C \to X$ is a contraction, then there exists a contraction $\tilde{T} : X \to X$ such that $\tilde{T}|_C = T$ and $\tilde{T}(X)$ is a subset of the closed convex hull of $T(C)$. $\tag{13.2}$

Condition (13.2) for X and C is easily seen to be equivalent to the following:

there exists a *contractive retraction* $P : X \to C$ (i.e. there exists a contraction P such that $P(X) = C$ and $P^2 = P$). $\tag{13.3}$

It is evident from the proof of Theorem 11.3 that (13.2) holds if X is a Hilbert space and C is any subset of X. As in other situations, (13.2) may be used to distinguish Euclidean spaces among a large class of finite-dimensional Banach spaces. The following theorem gives sufficient conditions on X and C in order that (13.2) fail.

Theorem 13.2. *Let X be a real reflexive Banach space and C a closed convex subset of X containing 0 in its interior and such that*

> *C has a unique supporting hyperplane $\Pi + y_0$ at one of its boundary   (13.4) points y_0 and Π is not the range of a norm 1 linear projection.*

Then there exists no contractive retraction from X onto C.

The following result [41] will be used in the proof of Theorem 13.2.

Lemma 13.3. *Let X be a reflexive Banach space and Π a hyperplane in X which is not the range of a linear projection of norm 1. Then given $y_0 \notin \Pi$, there exists $x_1, \ldots, x_n \in \Pi$ such that*

$$(\bigcap_{j=1}^{n} B(x_j, \|y_0 - x_j\|)) \cap \Pi = \emptyset.$$

Proof. The sets $B(x, \|y_0 - x\|) \cap \Pi$ are weakly compact. If the lemma does not hold, they satisfy the finite intersection property and there exists $y^* \in (\bigcap_{x \in \Pi} B(x, \|y_0 - x\|)) \cap \Pi$. Each $z \in X$ can be written uniquely as $z = x + \alpha y_0$ where $x \in \Pi$ and $\alpha \in \mathbb{R}$. Let $P(z) = x + \alpha y^*$. Then P is a linear map of X onto Π and $P^2 = P$. If $z \in \Pi$, then $\|Pz\| = \|z\|$. If $z = x + \alpha y_0$, $\alpha \neq 0$, then

$$\|P(z)\| = \|x + \alpha y^*\| = |\alpha| \, \| -\frac{x}{\alpha} - y^* \|$$

$$\leq |\alpha| \, \| -\frac{x}{\alpha} - y_0 \| = \|x + \alpha y_0\| = \|z\|.$$

Thus $\|P\| = 1$ contrary to the hypothesis of the lemma. $\square$

Proof of Theorem 13.2. Suppose C, y_0 and Π are as in (13.4). By Lemma 13.3 there exists $n \geq 1$ and points $x_1, \ldots, x_n \in \Pi$ such that

$$\bigcap_{j=1}^{n} B(x_j, \|y_0 - x_j\|)) \cap \Pi = \emptyset.$$

Let $f \in X^*$ such that $\Pi = \{x : f(x) = 0\}$ and $f(y_0) = 1$. Since $K = \bigcap_{j=1}^{n} B(x_j, \|y_0 - x_j\|)$ is weakly compact and nonempty, f achieves its minimum on K at some point $w_0 = \delta_0 y_0 + x_0$, $x_0 \in \Pi$, and since $K \cap \Pi = \emptyset$, $f(w_0) = \delta_0 > 0$. Let $z_0 = \delta_0 y_0$ and $y_j = x_j - x_0$, $1 \leq j \leq n$. If

$$z \in \bigcap_{j=1}^{n} B(y_j, \|z_0 - y_j\|), \text{ then } \|z - (x_j - x_0)\| \leq \|\delta_0 y_0 - (x_j - x_0)\|, \text{ or}$$

$\|(z + x_0) - x_j\| \leq \|w_0 - x_j\| \leq \|y_0 - x_j\|$ and $z + x_0 \in K$. Therefore

> if $z = \gamma y_0 + x$ $(x \in \Pi)$ and $z \in \bigcap_{j=1}^{n} B(y_j, \|z_0 - y_j\|)$, then $\gamma \geq \delta_0$.   (13.5)

For each λ, $0 \leq \lambda \leq 1$, let $w_\lambda = (1 + \lambda \delta_0) y_0$ and $z_{j,\lambda} = y_0 + \lambda y_j$ for $1 \leq j \leq n$.

If $z = \mu y_0 + x$ $(x \in \Pi)$ and $z \in \bigcap\limits_{j=1}^{n} B(z_{j,\lambda}, \|w_\lambda - z_{j,\lambda}\|)$, then $\qquad$ (13.6)

$$\|z - y_0 - \lambda y_j\| \leq \|\lambda(\delta_0 y_0 - y_j)\| \quad (1 \leq j \leq n) \text{ and}$$

$$\frac{z - y_0}{\lambda} \in \bigcap\limits_{j=1}^{n} B(y_j, \|z_0 - y_j\|).$$

By (13.5), with z as in (13.6), we have

$$(\mu - 1)/\lambda \geq \delta_0 \quad \text{or} \quad \mu \geq 1 + \delta_0 \lambda. \qquad (13.7)$$

Let ρ be the Minkowski functional on C, that is $\rho(x) = \inf\{\beta > 0 : x \in \beta C\}$. Then $\rho(y) \geq 1$ if $y \in \Pi + y_0$ since $\Pi + y_0$ is tangent to C at y_0. With z as in (13.6) it follows from (13.7) that

$$\rho(z) = \rho(\mu y_0 + x) = \mu \rho(y_0 + x/\mu) \geq \mu \geq \lambda \delta_0 + 1 = \rho(w_\lambda). \qquad (13.8)$$

Define the real-valued functions $h(\lambda) = \rho(w_\lambda) = 1 + \delta_0 \lambda$ and $g_j(\lambda) = \rho(z_{j,\lambda}) = \rho(y_0 + \lambda y_j)$, $0 \leq \lambda \leq 1$, $1 \leq j \leq n$. Clearly $h(0) = g_j(0) = 1$ and $h'(0) = \delta_0 > 0$. The existence of a unique supporting hyperplane at y_0 is equivalent to the Gâteaux differentiability of the Minkowski functional ρ at y_0 [49]. Thus, if D is the Gâteaux differential, then $g_j'(0) = D(\rho(y_0))(y_j) = 0$ since $D(\rho(y_0))(\Pi) = 0$ and $y_j \in \Pi (1 \leq j \leq n)$ Thus there exists λ^*, $0 < \lambda^* < 1$, such that

$$0 < \rho(z_{j,\lambda^*}) = g_j(\lambda^*) < h(\lambda^*) = \rho(w_{\lambda^*}) \quad (1 \leq j \leq n).$$

Let $\alpha = \max\limits_{1 \leq j \leq n} \{g_j(\lambda^*)\}$ and put $u_j = \frac{1}{\alpha} z_{j,\lambda^*}$ $(1 \leq j \leq n)$ and $v = \frac{1}{\alpha} w_{\lambda^*}$. Observe that

$$\rho(u_j) = \frac{1}{\alpha}\rho(z_{j,\lambda^*}) = \frac{1}{\alpha}g_j(\lambda^*) \leq 1 \ (1 \leq j \leq n), \text{ but } \rho(v) = \frac{1}{\alpha}\rho(w_{\lambda^*}) > \frac{1}{\alpha}g_j(\lambda^*) \ (1 \leq j \leq n).$$

Thus $u_j \in C$ $(1 \leq j \leq n)$ and $v \notin C$.

Suppose P is a contractive retraction of X onto C and let $z = Pv$. Then $\|z - u_j\| = \|Pv - Pu_j\| \leq \|v - u_j\|$ $(1 \leq j \leq n)$ and thus

$$\left\| z - \frac{1}{\alpha} z_{j,\lambda^*} \right\| \leq \left\| \frac{1}{\alpha} w_{\lambda^*} - \frac{1}{\alpha} z_{j,\lambda^*} \right\| \quad (1 \leq j \leq n).$$

Hence

$$\|\alpha z - z_{j,\lambda^*}\| \leq \|w_{\lambda^*} - z_{j,\lambda^*}\| \quad (1 \leq j \leq n) \text{ and}$$

$$\alpha z \in \bigcap\limits_{j=1}^{n} B(z_{j,\lambda^*}, \|w_{\lambda^*} - z_{j,\lambda^*}\|).$$

Then, by (13.8), $\rho(\alpha z) \geq \rho(w_{\lambda^*})$ and $\rho(z) \geq \frac{1}{\alpha}\rho(w_{\lambda^*}) = \rho(v) > 1$, contrary to the assumption $z \in C$. $\square$

Theorem 13.4. *Suppose X is a strictly convex real Banach space with $3 \leq \dim X < \infty$ and such that there exists a unique supporting hyperplane at each point of the unit ball of X (if $\dim X = 3$, strict convexity may be omitted). Then there exists a contractive retraction over the unit ball of X if and only if X is a Euclidean space.*

Proof. Suppose X has dimension n and X is not a Euclidean space. We will show (13.4) is satisfied with C the unit ball of X. If Π is any subspace of X of dimension $n-1$, then $\Pi + y_0$ is tangent to C at any point $y_0 \in C$ such that $\|y_0\| = 1$ and y_0 is at a maximum distance from Π. Thus the proof will be complete provided there is an $(n-1)$-dimensional subspace which is not the range of a linear projection of norm 1. Kakutani [41] characterized Hilbert space as the only Banach space in which every two dimensional subspace is the range of a linear projection of norm 1. This completes the proof if $n = 3$.

Suppose $n > 3$ and X is strictly convex. We will show that if Π_1 and Π_2 are the ranges of linear projections of norm 1, then so is $\Pi_1 \cap \Pi_2$. Thus, if every $(n-1)$-dimensional subspace of X is the range of a linear projection of norm 1, so is every two-dimensional subspace and X is a Euclidean space. Let Π_1 and Π_2 be subspaces of X and P_1 and P_2 norm 1 linear projections of X onto Π_1 and Π_2, respectively. For $x_0 \in X$ consider the sequence $\{x_m\}_{m=1}^\infty$ defined by

$$x_m = \frac{1}{m}\sum_{i=1}^m (P_1 P_2)^i x_0.$$

Since $\|(P_1 P_2)^i\| \leq 1$ ($i \geq 1$), we have $\|x_m\| \leq \|x_0\|$ for $m \geq 1$ and therefore the sequence $\{x_m\}_{m=1}^\infty$ has a convergent subsequence. Then by the Kakutani-Yosida mean ergodic theorem [92], the sequence $\{x_m\}_{m=1}^\infty$ converges to some $x \in X$. Clearly $P_1 P_2 x = x$ and therefore $x \in \Pi_1$. If $\|P_2 x\| = \|x\|$, then $P_2(\frac{1}{2}(x - P_2 x) + P_2 x) = P_2(x)$, so that

$$\|P_2 x\| \leq \|\tfrac{1}{2}(x - P_2 x) + P_2 x\| = \|\tfrac{1}{2}(x + P_2 x)\| \leq \tfrac{1}{2}(\|x\| + \|P_2 x\|) = \|P_2 x\|.$$

Thus, by the strict convexity of x, $P_2 x = x$. Hence if $x \notin \Pi_2$, then $\|P_2 x\| < \|x\|$ and $\|P_1 P_2 x\| < \|x\|$. Thus $x \in \Pi_1 \cap \Pi_2$ and the map Q defined by

$$Q(x_0) = \lim_{m \to \infty} \frac{1}{m}\sum_{i=1}^m (P_1 P_2)^i x_0 \quad (x_0 \in X)$$

is a linear norm 1 projection of X onto $\Pi_1 \cap \Pi_2$. $\square$

By restricting the range of the extension, B. Grünbaum [32] was able to eliminate the spaces with the binary intersection property in Theorem 12.11 when dim $X > 2$.

Theorem 13.5. *Let X be a Banach space. Every contraction $T : S \to X$ can be extended to all of X with range in the closed convex hull of $T(S)$ if and only if X is a Hilbert space or X is two-dimensional and has the binary intersection property.*

Proof. Suppose contractions on X always extend as described above. Then the added restriction on the range of the extension forces (Y, Y) to have the contraction extension property for every closed subspace Y of X. If dim $X \leq 2$, the implication follows from Theorem 12.11, so we assume that dim $X > 2$. If Y is any two-dimensional subspace of X, it follows from Theorem 12.11 that the unit ball of Y is either a parallelogram or an ellipse, depending on the strict convexity of Y. Obviously one cannot continuously transform an ellipse centered at the origin in $\mathbb{R}^n$ into a parallelogram centered at the origin using only figures of these types bounded away from 0. Thus, either X is a Hilbert space, or every two-dimensional subspace has unit ball a parallelogram. We show the latter situation cannot occur when dim $X > 2$.

Suppose Y is a finite-dimensional subspace of X, dim $Y > 2$, and every two-dimensional subspace of Y has unit ball a parallelogram. Let F be a two-dimensional face on the surface of the unit ball of Y and suppose L_1 and L_2 are two nonparallel line segments in the boundary of F which intersect at y_0. Such an F exists because of (12.15) and the Krein-Milman theorem. By choosing points $x_1 \in L_1$ and $x_2 \in L_2$ near y_0 and considering the two-dimensional subspace they generate, we see that the distance from 0 to the boundary of the unit ball in Y is 0, a contradiction.

The converse follows from the proof of Theorem 11.3 if X is a Hilbert space, and from elementary considerations if X is two-dimensional with unit ball a parallelogram. $\square$

Valentine [86] investigated the contraction extension property for (X,X) with X the surface of the unit ball in a Euclidean space. Although the space is not linear, it is embedded in a linear space and there is a notion of convexity.

Let S be the surface of the unit ball in $\mathbb{R}^n$, Π a hyperplane through 0, and S' one of the hemispheres determined by Π. Suppose d is a metric on S which is either the Euclidean or spherical metric.

Definition 13.6. *For $U \subseteq S'$ at a positive distance from Π, let*

$$\hat{U} = \{u \in \mathbb{R}^n : u = \sum_{i=1}^{m} \alpha_i u_i, \alpha_i > 0, u_i \in U \text{ and } m = 1,2,\dots\},$$

let $\tilde{co}(\hat{U}) = \hat{U} \cap S'$, and say U is spherically convex if $U = \tilde{co}(\hat{U})$.

Before Valentine's results we give three lemmas on spherical convexity. Lemma 13.8 can be found in [46]. In each lemma S, S' and Π are as above.

Lemma 13.7. *If U and V are spherically convex, closed, disjoint subsets of S'/Π, then there is a hyperplane Π' through 0 separating U and V.*

Proof. Let $\varepsilon = \min\{d(U,\Pi), d(V,\Pi)\}$ and $\delta = \min\{\varepsilon, d(coU, coV)\}$, where coU denotes the convex hull of U in $\mathbb{R}^n$. Choose $u_0 \in U$ and let A be the union of U and the ball in $\mathbb{R}^n$ of radius $\delta/2$ about u_0. For $x \in coA$, write

$$x = \sum_{i=1}^{r} \alpha_i u_i + \sum_{j=r+1}^{m} \alpha_j b_j \text{ where } u_i \in U \ (1 \leq i \leq r), \|b_j - u_0\| \leq \delta/2$$

$(r+1 \leq j \leq m)$ and $\sum_{i=1}^{m} \alpha_i = 1$, $\alpha_i \geq 0 \ (1 \leq i \leq m)$. Then

$$\left(\sum_{i=1}^{r} \alpha_i u_i + \sum_{j=r+1}^{m} \alpha_j u_0\right) \in coU \text{ and}$$

$$\left\| x - \left(\sum_{i=1}^{r} \alpha_i u_i - \sum_{j=r+1}^{m} \alpha_j u_0\right)\right\| \leq \sum_{j=r+1}^{m} \alpha_j \|b_j - u_0\| \leq \delta/2 < \delta.$$

Since $(A \cap S')\hat{}$ consists of all positive multiples of $co(A \cap S')$, $(A \cap S')\hat{}$ is a convex set, with interior, disjoint from the convex set $\hat{V}$. Thus there is a hyperplane Π' in

$\mathbb{R}^n$ separating $(A \cap S')\hat{}$ and $\hat{V}$. Clearly Π' is through the origin and Π' separates U and V since $U \subseteq (A \cap S')\hat{}$. $\square$

Lemma 13.8. *Suppose $C_0,\dots,C_k$ are closed spherically convex subsets of $S' \setminus \Pi$, each k of which have a point in common. If $\bigcup\limits_{i=0}^{k} C_i$ is spherically convex then*
$$\bigcap\limits_{i=0}^{k} C_i \neq \emptyset.$$

Proof. If $k = 0$ or 1 the lemma is obvious; thus suppose the result is true for k sets, $k > 1$, and $(k+1)$ sets $C_0,\dots,C_k$ are given. If $\bigcap\limits_{i=0}^{k} C_i = \emptyset$ then, by Lemma 13.7, there is a hyperplane Π' through 0 separating the nonempty spherically convex sets C_0 and $D = \bigcap\limits_{i=1}^{k} C_i$. Let $D_i = C_i \cap \Pi'$ $(1 \le i \le k)$. Since $C_i \cap C_0 \neq \emptyset$, $C_i \cap D \neq \emptyset$ and C_i is connected $(1 \le i \le k)$, it follows that each D_i is nonempty. Furthermore each D_i is closed and spherically convex and $\bigcup\limits_{i=1}^{k} D_i = \bigcup\limits_{i=1}^{k} (C_i \cap \Pi') = (\bigcup\limits_{i=0}^{k} C_i) \cap \Pi'$, so $\bigcup\limits_{i=1}^{k} D_i$ is spherically convex. For $1 \le j \le k$, $\bigcap\limits_{\substack{i=1 \\ i \neq j}}^{k} D_i = (\bigcap\limits_{\substack{i=1 \\ i \neq j}}^{k} C_i) \cap \Pi'$ and $\bigcap\limits_{\substack{i=1 \\ i \neq j}}^{k} C_i$ is a spherically convex set which intersects both C_0 and D, and therefore intersects Π'. Then, by the inductive hypothesis $\bigcap\limits_{i=1}^{k} D_i \neq \emptyset$, but
$$\bigcap\limits_{i=1}^{k} D_i = (\bigcap\limits_{i=1}^{k} C_i) \cap \Pi' = D \cap \Pi' = \emptyset, \text{ a contradiction.} \quad \square$$

Lemma 13.9. *Let $a_0,\dots,a_r \in S' \setminus \Pi$ and let $A_0,\dots,A_r$ be closed spherically convex subsets of $S' \setminus \Pi$ such that $\tilde{co}\{a_{i_0},\dots,a_{i_s}\} \subseteq A_{i_0} \cup \dots \cup A_{i_s}$ for every subset $\{i_0,\dots,i_s\}$ of $\{0,1,\dots,r\}$. Then $\bigcap\limits_{i=0}^{r} A_i \neq \emptyset$.*

Proof. Again we induct on the number of sets. The lemma clearly holds for $r = 0,1$, so suppose the result is true for r sets and points $(r > 1)$ and $r+1$ sets are given. We may assume, without loss of generality, that $A_i \subseteq \tilde{co}\{a_0,\dots,a_r\}$ $(0 \le i \le r)$. Then by the inductive hypothesis, any r of the A_i's intersect and since $\bigcup\limits_{i=0}^{r} A_i = \tilde{co}\{a_0,\dots,a_r\}$, the lemma follows from Lemma 13.8. $\square$

The extension of contractions from S to S falls into two categories: maps with range in a hemisphere and maps whose range is contained in no hemisphere. In each case the key to the argument is the simple observation that

> if $x,y,z,w \in S$ and $d(x,y) \le d(z,w)$ $(d(x,y) < d(z,w))$, then $\quad$ (13.9)
> $(x,y) \ge (z,w)$ $((x,y) > (z,w))$ where (x,y) is the usual inner product in $\mathbb{R}^n$.

No matter which metric d is, $d(x,y) \le d(z,w)$ implies $\|x-y\|^2 \le \|z-w\|^2$ or
$$(x,x) - 2(x,y) + (y,y) \le (z,z) - 2(z,w) + (w,w).$$

All points of S have norm 1, so (13.9) follows.

Lemma 13.10. *Suppose S is the surface of the unit ball in $\mathbb{R}^n$ and S' is a hemisphere in $\mathbb{R}^k$. If T is a contraction from a subset of S into S', then T can be extended to a contraction $\tilde{T}: S \to S'$.*

Proof. As in other cases it suffices to suppose that the domain of T is finite and show extension to one more point is possible. Suppose $x_1,\ldots,x_m \in S$ and $y_1,\ldots,y_m \in S'$ such that $d(y_i,y_j) \le d(x_i,x_j)$ $(1 \le i, j \le m)$. For $x \in S$ we must find $y \in S'$ such that $d(y,y_j) \le d(x,x_j)$ $(1 \le j \le m)$. Thus we must show $\bigcap_{j=1}^m B_j \ne \emptyset$ where B_j is the closed ball in S' about y_j of radius $d(x,x_j)$. Let Π be the hyperplane in $\mathbb{R}^k$ which determines S'. We may assume none of the $y_j \in \Pi$, for Lipschitz conditions will be preserved by moving small distances along great circles. If there is nonempty intersection in these cases, a limit argument will show $\bigcap_{j=1}^m B_j \ne \emptyset$.

By Lemma 13.9 it suffices to show $\tilde{co}\{y_1,\ldots,y_m\} \subseteq \bigcup_{j=1}^m B_j$.

Suppose $y \in \tilde{co}\{y_1,\ldots,y_m\} \setminus (\bigcup_{j=1}^m B_j)$. Then $y = \sum_{j=1}^m \lambda_j y_j$ with $\lambda_j \ge 0$ $(1 \le j \le m)$ and $\sum_{i=1}^m \lambda_i > 0$. Since $y \notin B_i$ we have $d(y,y_i) > d(x,x_i)$ and hence, by (13.9),

$$(y,y_i) < (x,x_i) \qquad (1 \le i \le m). \tag{13.10}$$

By assumption $d(y_i,y_j) \le d(x_i,x_j)$, so (13.9) yields

$$(y_i,y_j) \ge (x_i,x_j) \qquad (1 \le i, j \le m). \tag{13.11}$$

It follows from (13.10) that

$$1 = (y,y) = \sum_{i=1}^m \lambda_i(y,y_i) < \sum_{i=1}^m \lambda_i(x,x_i) \le \left\| \sum_{i=1}^m \lambda_i x_i \right\|,$$

however from (13.11) we obtain

$$\left\| \sum_{i=1}^m \lambda_i x_i \right\|^2 = \sum_{i,j=1}^m \lambda_i \lambda_j (x_i,x_j) \le \sum_{i,j=1}^m \lambda_i \lambda_j (y_i,y_j) = \|y\|^2 = 1.$$

This contradiction completes the proof. $\square$

Before considering the second type of contraction we investigate the significance of being contained in no hemisphere.

Lemma 13.11. *If D is a subset of the surface S of the unit ball in $\mathbb{R}^n$, D is contained in no hemisphere of S and $x_0 \in D$, then there exists a positive integer m, points $x_1,\ldots,x_m \in D$, and positive numbers $\lambda_0,\ldots,\lambda_m$ such that $0 = \Sigma_{i=0}^m \lambda_i x_i$.*

Proof. Every hyperplane through 0 must intersect the closed convex hull of D, $\overline{coD}$. Thus $0 \in \overline{coD}$ and $\overline{coD}$ has no supporting hyperplane through 0. It then follows that 0 is an interior point of coD and the interior of the line segment from $-x_0$ to 0 must intersect coD. The lemma now follows trivially. $\square$

Lemma 13.12. *If T is a contraction from a subset of the surface S of the unit ball in $\mathbb{R}^n$ into S whose range is contained in no hemisphere, then T is an isometry and extends to an isometry of S onto S.*

Proof. Suppose x_0 is in the domain of T and let m, $T(x_1),\ldots,T(x_m)$ and $\lambda_0,\lambda_1,\ldots,\lambda_m$ be as in Lemma 13.11. Let $y_i = Tx_i$ $(0 \leq i \leq m)$ and suppose $d(y_0,y_1) < d(x_0,x_1)$. For any i we have either

$$\pi = d(x_0,x_i) + d(x_i,-x_0) = d(y_0,y_i) + d(y_i,-y_0)$$

or

$$4 = (d(x_0,x_i))^2 + (d(x_i,-x_0))^2 = (d(y_0,y_i))^2 + (d(y_i,-y_0))^2.$$

Thus, in any case, we have $d(y_i,-y_0) \geq d(x_i,-x_0)$ $(2 \leq i \leq m)$ and $d(y_1,-y_0) > d(x_1,-x_0)$. It follows from (13.9) that $(y_1,-y_0) < (x_1,-x_0)$ and $(y_i,-y_0) \leq (x_i,-x_0)$ $(2 \leq i \leq m)$. Hence

$$1 = (-y_0,-y_0) = \sum_{i=1}^{m} \frac{\lambda_i}{\lambda_0} (y_i,-y_0) < \sum_{i=1}^{m} \frac{\lambda_i}{\lambda_0} (x_i,-x_0) \leq \left\| \sum_{i=1}^{m} \frac{\lambda_i}{\lambda_0} x_i \right\|. \quad (13.12)$$

On the other hand $d(y_i,y_j) \leq d(x_i,x_j)$ implies that $(x_i,x_j) \leq (y_i,y_j)$ $(1 \leq i, j \leq m)$ and

$$\left\| \sum_{i=1}^{m} \frac{\lambda_i}{\lambda_0} x_i \right\|^2 = \sum_{i,j=1}^{m} \frac{\lambda_i \lambda_j}{\lambda_0 \lambda_0} (x_i,x_j) \leq \sum_{i,j=1}^{m} \frac{\lambda_i \lambda_j}{\lambda_0 \lambda_0} (y_i,y_j) = \|y_0\|^2 = 1. \quad (13.13)$$

This is a contradiction, so $d(y_0,y_1) = d(x_0,x_1)$ and $d(y_0,y_i) = d(x_0,x_i)$ $(1 \leq i \leq m)$. Even without the assumption in (13.12) that $d(y_0,y_1) < d(x_0,x_1)$, we obtain $1 \leq \left\| \sum_{i=1}^{m} (\lambda_i/\lambda_0)x_i \right\|$, so if $d(y_i,y_j) < d(x_i,x_j)$ for some i,j $(1 \leq i, j \leq m)$, we obtain a contradiction in (13.13). Therefore $d(x_i,x_j) = d(y_i,y_j)$ for $0 \leq i, j \leq m$.

There is a linear isometry of $\mathbb{R}^n$ onto $\mathbb{R}^n$ carrying x_i to y_i $(0 \leq i \leq m)$, so we may assume $x_i = y_i$ for all i. If x is in the domain of T and $d(Tx,Tx_0) < d(x,x_0)$, then $(Tx,x_0) > (x,x_0)$ and $(Tx,x_i) \geq (x,x_i)$ $(1 \leq i \leq m)$. Thus $\sum_{i=0}^{m} \lambda_i(Tx,x_i) > \sum_{i=0}^{m} \lambda_i(x,x_i)$, which is a contradiction since $\sum_{i=0}^{m} \lambda_i x_i = \sum_{i=0}^{m} \lambda_i y_i = 0$. Therefore $d(Tx,Tx_0) = d(x,x_0)$ and since x and x_0 were arbitrary elements in the domain of T, it follows that T is an isometry. Setting $T(0)=0$ and applying Theorem 11.4, we obtain the desired result. $\square$

The following is a summary of the results of Lemma 13.10 and 13.12.

Theorem 13.13. *Let S_1 and S_2 be Euclidean spheres both with the spherical or both with the Euclidean metric. Then (S_1,S_2) has the contraction extension property. If $S = S_1 = S_2$, $D \subset S$ and $T:D\to S$ is a contraction whose range is contained in no hemisphere, then the extension is an isometry. If S is the unit sphere in Hilbert space, then (S,S) does not have the contraction extension property.*

Proof. Only the last statement requires proof. For a counterexample (see Theorem 11.4), let S be the unit sphere in ℓ^2, $D = \{(0,x_1,x_2,\ldots)\in S\}$ and define $T:D\to S$ by $T((0,x_1,x_2,\ldots)) = (x_1,x_2,\ldots)$. Then T cannot be extended to the point $u = (1,0,0,\ldots)$ as a contraction on S. Assume that $Tu = (a_1,a_2,\ldots)$ with $\sum a_i^2 = 1$ and choose an integer j so that $a_j \neq 0$. Let $x\in D$ be such that $x_{j+1} = 1$ if $a_j < 0$ and $x_{j+1} = -1$ if $a_j > 0$ and $x_i = 0$ otherwise. Then $\|u - x\|^2 = 2$, but

$$\|Tu - Tx\|^2 = (1 + |a_j|)^2 + \sum_{i \neq j} a_i^2 = 2 + 2|a_j| > 2. \quad \square$$

We conclude this section with a brief discussion of the problem of extending uniformly continuous mappings. Suppose that X and Y are metric spaces, $E \subset X$

and $T: E \to Y$. The *modulus of continuity* for T is the function δ_T defined on $\mathbb{R}^+$ by

$$\delta_T(t) = \sup\{d_2(Tx_1, Tx_2): x_1, x_2 \in E \text{ and } d_1(x_1, x_2) \leq t\}.$$

It is clear that δ_T is a nonnegative and nondecreasing function and that T is uniformly continuous on E if and only if $\lim_{t \to 0} \delta_T(t) = 0$. It is also evident that δ_T is the smallest function δ such that

$$d_2(Tx_1, Tx_2) \leq \delta(d_1(x_1, x_2)) \qquad (x_1, x_2 \in E). \tag{13.14}$$

Any nonnegative function δ on $\mathbb{R}^+$ for which (13.14) holds is termed a modulus of continuity for T.

If T has a uniformly continuous extension T' to all of X and $\delta_{T'}$ is its modulus of continuity, then $\delta_T(t) \leq \delta_{T'}(t)$ for $t \in \mathbb{R}^+$. If it happens that X is metrically convex (see §10), then δ_T is subadditive. So, for example, if X is a linear space and T is a uniformly continuous map from a subset of X into a metric space Y, then in order that T admit a uniformly continuous extension to all of X, it is necessary that δ_T be majorized by a nondecreasing and subadditive function that tends to zero as $t \to 0$. In case T is a contraction or a Lipschitz-Hölder map, these conditions are automatically met since it has a modulus of continuity of the form $\delta(t) = kt^\alpha (0 < \alpha \leq 1)$; and extension is always possible, according to Theorem 11.2, provided that Y is metrically convex and has the binary intersection property. The more general situation described in the following theorem includes results of McShane [56] when $Y = \mathbb{R}$ and of Aronszajn and Panitchpadki [3] on general spaces with the binary intersection property.

Theorem 13.14. *Let Y be a metric space which is metrically convex and has the binary intersection property, and let δ be a nonnegative, nondecreasing and subadditive function on $\mathbb{R}^+$ such that $\delta(0) = \lim_{t \to 0} \delta(t) = 0$. Then for any metric space X, $E \subset X$ and map $T: E \to Y$ satisfying (13.14), there exists an extension of T to all of X preserving (13.14).*

Proof. We may assume $\delta(t) > 0$ for $t > 0$, otherwise only constant maps satisfy (13.14). Then $\delta \circ d_1$ is a metric on X and the problem of extending T reduces to the contraction extension problem for the pair (X, Y), X having the metric $\delta \circ d_1$. The conclusion now follows from Theorem 11.2. $\square$

In case $Y = \mathbb{R}$ it is possible to give an explicit construction for the extension. The idea is due to McShane [56]. Also see [17].

Lemma 13.15. *Suppose δ_0 is a nonnegative function on $\mathbb{R}^+$ such that $\delta_0(t) \leq ht + k$ for some constants h and k. Then there is a continuous and concave function δ on $\mathbb{R}^+$ such that $\delta_0(t) \leq \delta(t)$ for $t \geq 0$. Moreover, whenever $\delta_0(t)$ tends to zero as $t \to 0$, δ can be chosen to do the same.*

Proof. Let $Q = \{(t, u): 0 \leq t \text{ and } u \leq \delta_0(t)\}$. By hypothesis Q is contained in a half-plane that does not contain the first quadrant. Hence the intersection of all half-planes containing Q is a convex set whose upper boundary is a concave curve $\delta(t)$ for $t \geq 0$, and $\delta(t) \geq \delta_0(t)$ for $t \geq 0$.

Suppose $\lim_{t \to 0} \delta_0(t) = 0$. Then for every $\varepsilon > 0$ there exists a $d > 0$ such that

$0 < \delta_0(t) < \varepsilon$ for $0 < t \leq d$. For positive values of a, $at + \varepsilon \geq \delta_0(t)$ for $0 < t \leq d$, and, if a is sufficiently large, $at + \varepsilon \geq ht + k$ for $t \geq d$. Hence $0 \leq \limsup_{t \to 0} \delta(t) \leq \varepsilon$. As this is true for every positive ε, the limit must be 0. $\square$

Note that a nonnegative and concave function on $\mathbb{R}^+$ is necessarily nondecreasing and subadditive.

Theorem 13.16. *Let X be a metric space, $E \subset X$ and $T : E \to \mathbb{R}$ a map which has a modulus of continuity δ that is nondecreasing, subadditive and for which $\lim_{t \to 0} \delta(t) = 0$. Then the map T' defined at each $x \in X$ by*

$$T'(x) = \sup\{T(t) - \delta(d(t,x)) : t \in E\} \tag{13.15}$$

is an extension of T to X which preserves the modulus of continuity δ.

Proof. For every $x \in E$ we have $T'(x) = T(x)$ since, by hypothesis, $T(t) - \delta(d(t,x)) \leq T(x)$ for all $t \in E$, and the upper bound is attained when $t = x$. Since δ is nondecreasing, $|T(t) - T(s)| \leq \delta(d(t,s)) \leq \delta(d(t,x)) + \delta(d(s,x))$ or $T(t) - \delta(d(t,x)) \leq T(s) + \delta(d(s,x))$. Hence, fixing s in E and x in X, and letting t vary over E, we see that $T'(x) < \infty$. Now let x_1 and x_2 be any two points in X and suppose that $T'(x_1) \leq T'(x_2)$. Then, using the fact that δ is nondecreasing and subadditive, we have

$$0 \leq T'(x_2) - T'(x_1) = \sup\{T(t) - \delta(d(t,x_2)) : t \in E\} - \sup\{T(t) - \delta(d(t,x_1)) : t \in E\}$$

$$\leq \sup\{[T(t) - \delta(d(t,x_2))] - [T(t) - \delta(d(t,x_1))] : t \in E\}$$

$$\leq \sup\{\delta(d(t,x_2) + d(x_1,x_2)) - \delta(d(t,x_2)) : t \in E\}$$

$$\leq \delta(d(x_1,x_2)). \quad \square$$

Corollary 13.17. *If T is a bounded and uniformly continuous map from a subset E of X into $\mathbb{R}$, then T has a uniformly continuous extension to X which preserves the bounds on T.*

Proof. The modulus of continuity of T is bounded and tends to zero as $t \to 0$. Hence, by Lemma 13.15, there exists a modulus of continuity δ for T which is concave, tends to zero as $t \to 0$ and satisfies $\delta_T(t) \leq \delta(t)$ for $t \geq 0$. By the previous result T has an extension T' to all of X with the same modulus of continuity δ. It is clear from (13.15) that T' has the same upper bound as T. If m is the lower bound for T, define T'' by $T''(x) = T'(x)$ when $T'(x) \geq m$ and $T''(x) = m$ when $T'(x) < m$. Then T'' has the same bounds as T, is an extension of T, and has δ as a modulus of continuity. $\square$

A different approach is needed to establish results of this type in case the range space Y does not have the binary intersection property. E. Mickle [59] was the first to notice that the quadratic forms (2.9), used by Schoenberg in connection with the isometric embedding problem, are also central to this class of extension problems. We shall restrict our attention to Hilbert space.

Suppose H is a Hilbert space and $\delta : \mathbb{R}^+ \to \mathbb{R}^+$ satisfies the conditions $\delta(0) = 0$ and

$$\sum_{i,j=1}^{n} \{\delta^2(\|x_i - x_0\|) + \delta^2(\|x_j - x_0\|) - \delta^2(\|x_i - x_j\|)\} \xi_i \xi_j \geq 0 \qquad (13.16)$$

for any finite set $x_0, x_1, \ldots, x_n$ in H and nonnegative numbers $\xi_1, \xi_2, \ldots, \xi_n$.

This is a weaker condition than (2.9), at least in infinite-dimensional L^p spaces for $2 < p < \infty$. This is a consequence of (15.2) and Theorem 8.1, that is, $\delta(t) = t^{2\alpha}$ ($1 \leq 2\alpha \leq p'$) satisfies (13.16) but not (2.9) since only the zero function belongs to $N(L^p)$. It is of some interest to note that any δ satisfying (13.16) induces a semi-metric on H. To see this, choose three points $x_0, x_1, x_2 \in H$. From (13.16),

$$\{\delta^2(\|x_1 - x_0\|) + \delta^2(\|x_2 - x_0\|) - \delta^2(\|x_1 - x_2\|)\} \xi_1 \xi_2 +$$
$$\{\delta^2(\|x_1 - x_0\|) \xi_1^2 + \delta^2(\|x_2 - x_0\|) \xi_2^2 \geq 0,$$

or $\qquad\qquad\qquad\qquad\qquad\qquad\qquad\qquad\qquad\qquad\qquad (13.17)$

$$[\delta(\|x_1 - x_0\|) \xi_1 - \delta(\|x_2 - x_0\|) \xi_2]^2$$
$$+ \{[\delta(\|x_1 - x_0\|) + \delta(\|x_2 - x_0\|)]^2 - \delta^2(\|x_1 - x_2\|)\} \xi_1 \xi_2 \geq 0.$$

If $\delta(\|x_2 - x_0\|) = 0$, we have

$$\xi_1^2 \delta^2(\|x_1 - x_0\|) + \xi_1 \xi_2 \delta^2(\|x_1 - x_0\|) \geq \delta^2(\|x_1 - x_2\|) \xi_1 \xi_2.$$

Dividing this last expression by ξ_1 and then letting ξ_1 approach zero, we obtain $\delta^2(\|x_1 - x_0\|) \geq \delta^2(\|x_1 - x_2\|)$. Therefore, by symmetry,

$$\delta(\|x_1 - x_0\|) = \delta(\|x_1 - x_2\|) \quad \text{if} \quad \delta(\|x_2 - x_0\|) = 0. \qquad (13.18)$$

In particular,

$$\delta(\|x_1 - x_2\|) \leq \delta(\|x_1 - x_0\|) + \delta(\|x_0 - x_2\|) \qquad (13.19)$$

if one of the three quantities is zero. If none vanish, then (13.19) follows from (13.17) by choosing $\xi_1 = \delta(\|x_2 - x_0\|)$ and $\xi_2 = \delta(\|x_1 - x_0\|)$. Thus $\delta(\|x - y\|)$ is a semi-norm on H. Moreover, by (13.18), the relation $x \sim y$ if and only if $\delta(\|x - y\|) = 0$ is an equivalence relation and, therefore, $\delta(\|x - y\|)$ is a metric on $H/\sim$.

Lemma 13.18. *Suppose that δ satisfies (13.16), $E \subset H$ and $T : E \to H$ satisfies*

$$\|Tx - Ty\| \leq \delta(\|x - y\|) \qquad (x, y \in H). \qquad (13.20)$$

Then T has an extension to H which preserves (13.20). Moreover, the extension may be chosen so that its range lies in the closed convex hull of $T(E)$.

Proof. The proof is identical to that of Theorem 11.3 with $\delta^2(\|x_i - x_j\|)$ in place of $\|x_i - x_j\|^{2\alpha}$. $\square$

Lemma 13.19. *If δ satisfies (13.16), then so does $\delta_\lambda(t) = \min\{\delta(t), \lambda\}$ ($t \geq 0$) for each $\lambda > 0$.*

Proof. Divide the sum

$$\sum_{i,j=1}^{n} \left[\delta_\lambda^2(\|x_i - x_0\|) + \delta_\lambda^2(\|x_j - x_0\|) - \delta_\lambda^2(\|x_i - x_j\|) \right] \xi_i \xi_j$$

into two sums Σ' and Σ'', the first extending over all pairs i, j with $\max\{\delta(\|x_i - x_0\|),$

$\delta(\|x_j - x_0\|)\} \leq \lambda$, and the second extending over the remaining pairs. The first sum can be put in the form

$$\Sigma'_{i,j} \left[\delta^2(\|x_i - x_0\|) + \delta^2(\|x_j - x_0\|) - \delta^2(\|x_i - x_j\|) \right] \xi_i \xi_j$$

$$+ \; \Sigma'_{i,j} \left[\delta^2(\|x_i - x_j\|) - \delta_\lambda^2(\|x_i - x_j\|) \right] \xi_i \xi_j$$

and is therefore nonnegative. The second sum is also nonnegative because each of its terms is nonnegative. $\square$

Lemma 13.20. *If* δ *satisfies* (13.16) *and* K *is a nonnegative and concave function on* $\mathbb{R}^+$, *then* $\sigma(t) = [K(\delta^2(t))]^{1/2}$ *also satisfies* (13.16).

Proof. Assume first that K is twice continuously differentiable. Then K is nondecreasing and nonnegative, and K' is nonincreasing and nonnegative. Therefore $K(0+) = \lim_{u \to 0^+} K(u)$ and $K'(\infty) = \lim_{u \to \infty} K'(u)$ exist and are finite and nonnegative. Therefore

$$K(u) - K(0^+) = \int_0^u K'(s)ds = uK'(u) - \int_0^u sK''(s)ds$$

$$= uK'(\infty) - \int_0^u sK''(s)ds - \int_u^\infty uK''(s)ds$$

$$= uK'(\infty) - \int_0^\infty (\inf\{u,s\}) K''(s)ds.$$

Hence

$$\sigma(t)^2 = K(\delta^2(t)) = K(0^+) + K'(\infty)\delta^2(t) - \int_0^\infty (\delta_{\sqrt{s}}(t))^2 K''(s)ds.$$

Consequently,

$$\sum_{i,j \neq 0} \left[\sigma^2(\|x_i - x_0\|) + \sigma^2(\|x_j - x_0\|) - \sigma^2(\|x_i - x_j\|) \right] \xi_i \xi_j$$

$$= K(0^+) \left(\sum_{i=1}^n \xi_i \right)^2$$

$$+ K'(\infty) \sum_{i,j=1}^n \left[\delta^2(\|x_i - x_0\|) + \delta^2(\|x_j - x_0\|) + \delta^2(\|x_i - x_j\|) \right] \xi_i \xi_j$$

$$+ \int_0^\infty \sum_{i,j=1}^n \left[\delta^2_{\sqrt{s}}(\|x_i - x_0\|) + \delta^2_{\sqrt{s}}(\|x_j - x_0\|) - \delta^2_{\sqrt{s}}(\|x_i - x_j\|) \right] \xi_i \xi_j (-K''(s))ds]$$

$$\geq 0.$$

For the general case, let

$$K_\varepsilon(u) = \int_0^\infty K(u/v) \, \varphi_\varepsilon (\log v) \, dv/v,$$

where φ_ε is nonnegative C^∞ with support in $[-\varepsilon, \varepsilon]$ and $\int_{-\varepsilon}^\varepsilon \varphi_\varepsilon(t)dt = 1$. Clearly K_ε is concave and nonnegative. Also

$$K_\varepsilon(u) = \int_0^\infty K(w) \varphi_\varepsilon (\log u/w) dw/w$$

and therefore K_ε is a C^∞ function. It follows from the substitution $v = e^t$ that $\int_0^\infty \varphi_\varepsilon(\log v) dv/v = \int_{-\varepsilon}^{\varepsilon} \varphi_\varepsilon(t) dt = 1$. Therefore

$$|K_\varepsilon(u) - K(u)| = \left| \int_0^\infty [K(u/v) - K(u)] \, \varphi_\varepsilon(\log v) \, dv/v \right|$$

$$\leq \int_0^{1-\eta} |K(u/v) - K(u)| \, \varphi_\varepsilon(\log v) \, dv/v$$

$$+ \sup_{1-\eta \leq v \leq 1+\eta} |K(u/v) - K(u)| + \int_{1+\eta}^\infty |K(u/v) - K(u)| \varphi_\varepsilon(\log v) dv/v.$$

Now K is continuous so, as $\varepsilon \to 0$, $K(u) \to K(u)$ uniformly in any closed subinterval $[a,b]$ of $(0,\infty)$. $\square$

These results lead to the following theorem of F. Grünbaum and E. H. Zarantonello [33].

Theorem 13.21. *In order that a uniformly continuous mapping T from a subset of Hilbert space H into H have a uniformly continuous extension to all of H, it is necessary and sufficient that its modulus of continuity admit a nondecreasing and subadditive majorant δ_0 with $\delta_0(0) = \lim_{t \to 0} \delta_0(t) = 0$.*

Proof. As we have already pointed out, the modulus of continuity of a uniformly continuous extension of T to all of H is such a majorant.

Suppose T is uniformly continuous from a subset of H into H and δ_0 is a nondecreasing and subadditive majorant of its modulus of continuity. According to Lemma 13.15 there exists a nonnegative and concave δ that majorizes δ_0. A straightforward calculation shows that the function $K(t) = \delta(\sqrt{t})^2$ $(t \geq 0)$ is also concave in case δ has two continuous derivatives; and an argument similar to that of the preceding lemma shows that K is concave in general. Now the identity function certainly satisfies condition (13.16) on H, hence, by Lemma 13.20, $K(t^2)^{1/2} = \delta(t)$ also satisfies condition (13.16). The desired conclusion follows from Lemma 13.18. $\square$

Chapter IV. Interpolation and L^p Inequalities

In this chapter we develop a multi-component version of the Riesz-Thorin interpolation theorem and use it to derive a number of L^p inequalities which are natural relatives of inequality (4.15) and the now standard inequalities of Clarkson. These inequalities are crucial to the problem of extending Lipschitz-Hölder maps of order α between L^p spaces (see §19). In addition they are of considerable intrinsic interest, a point we here emphasize by applying certain of their number to a packing problem in L^p.

The interpolation result is due to Wells and Hayden [37], the inequalities come from [90], and our treatment of the packing problem is based on a paper of Burlak, Rankin and Robertson [14].

§14. A Multi-Component Riesz-Thorin Theorem

Let $(\Omega_1,\mu_1),(\Omega_2,\mu_2),\ldots,(\Omega_n,\mu_n)$ be σ-finite measure spaces and $P = (p_1,p_2,\ldots,p_n)$ an n-tuple with $1 \leq p_k \leq \infty$. On the direct sum $\bigoplus L^{p_k}(\mu_k)$, consisting of the vectors

$$x = (x_1,x_2,\ldots,x_n),\ x_k \in L^{p_k}(\mu_k), \tag{14.1}$$

equipped with the usual coordinate addition and scalar multiplication, introduce the weighted norm

$$\|x\|_{P,r} = \left[\sum_{k=1}^{n} \left(\int_{\Omega_k} |x_k|^{p_k}\, d\mu_k \right)^{r/p_k} \lambda_k \right]^{1/r}$$

$$= \left(\sum_{k=1}^{n} \|x_k\|_{p_k}^{r}\, \lambda_k \right)^{1/r}, \tag{14.2}$$

where $1 \leq r < \infty$ and $\lambda = (\lambda_1,\lambda_2,\ldots,\lambda_n)$ is an n-tuple of positive weights. In case $r = \infty$,

$$\|x\|_{P,\infty} = \max_{1 \leq k \leq n} \|x_k\|_{p_k}. \tag{14.3}$$

We denote the resulting space by $L^{P,r}(\lambda)$.

It is easy to show that $L^{P,r}(\lambda)$ is a Banach space. Further, if $p' = p/(p-1)$ denotes the index conjugate to p and $P' = (p'_1,p'_2,\ldots,p'_n)$, then

$$\int xy\, d\lambda = \sum_{k=1}^{n} \left(\int_{\Omega_k} x_k y_k\, d\mu_k \right) \lambda_k \qquad (y \in L^{P',r'}(\lambda)) \tag{14.4}$$

defines a bounded linear functional on $L^{P,r}(\lambda)$ of norm $\|y\|_{P',r'}$. It is a straight-forward verification to show that if $P < \infty$ (no component of P equals ∞), then $L^{P,r}(\lambda)$ is isomorphic and isometric to the Banach conjugate of $L^{P,r}(\lambda)$ and that (14.4) defines the form of a general linear functional on $L^{P,r}(\lambda)$. Furthermore, if $x \in L^{P,r}(\lambda)$, then

$$\|x\|_{P,r} = \sup \left| \int xy \, d\lambda \right|, \tag{14.5}$$

where y varies over all simple vectors in $L^{P',r'}(\lambda)$ of norm one. Here, and subsequently, a measurable vector is one having measurable components, and a simple vector is one whose components are simple measurable functions each vanishing outside some set of finite measure.

Corresponding to a second sequence $(N_1,v_1),(N_2,v_2),...,(N_m,v_m)$ of σ-finite measure spaces, define $L^{Q,s}(\eta)$ in an analogous way, where $Q = (q_1,q_2,...,q_m)$, $1 \le q_k \le \infty$, $1 \le s \le \infty$ and $\eta = (\eta_1,\eta_2,...,\eta_m)$ is an m-tuple of positive weights. In the following theorem T denotes a linear transformation with domain the simple vectors on $\Omega = (\Omega_1,\Omega_2,...,\Omega_n)$ and range in the measurable vectors on $N = (N_1,N_2,...,N_m)$.

Theorem 14.1. *Let*

$$1 \le P_i, Q_i \le \infty, 1 \le r_i \le \infty, 1 \le s_i \le \infty \quad \text{for } i=1,2$$

and

$$\frac{1}{P} = \frac{1-t}{P_1} + \frac{t}{P_2}, \frac{1}{Q} = \frac{1-t}{Q_1} + \frac{t}{Q_2}, \frac{1}{r} = \frac{1-t}{r_1} + \frac{t}{r_2}, \frac{1}{s} = \frac{1-t}{s_1} + \frac{t}{s_2},$$

where $0 \le t \le 1$, and assume there exist constants M_1 and M_2 such that

$$\|Tx\|_{Q_1,s_1} \le M_1 \|x\|_{P_1,r_1} \tag{14.6}$$

and

$$\|Tx\|_{Q_2,s_2} \le M_2 \|x\|_{P_2,r_2} \tag{14.7}$$

for any simple vector x on Ω. Then we may conclude that

$$\|Tx\|_{Q,s} \le M_1^{1-t} M_2^t \|x\|_{P,r}. \tag{14.8}$$

Furthermore, if $P < \infty$, T can be extended uniquely to $L^{P,r}(\lambda)$.

Proof. In explanation of our notation, the vectors

$$P_k = (p_{k1},p_{k2},...,p_{kn}), \quad Q_k = (q_{k1},q_{k2},...,q_{km}), \quad k=1,2,$$

have their components in $[1,\infty]$ and

$$P = (p_1,p_2,...,p_n), \quad Q = (q_1,q_2,...,q_m)$$

where

$$\frac{1}{p_k} = \frac{1-t}{p_{1k}} + \frac{t}{p_{2k}}, \quad k=1,2,...,n$$

$$\frac{1}{q_k} = \frac{1-t}{q_{1k}} + \frac{t}{q_{2k}}, \quad k=1,2,...,m.$$

Our proof closely follows that of the Riesz-Thorin theorem given in [93]. It follows from (14.5) and the linearity of T that (14.8) will hold provided we show that

$$\left|\int Txy\, d\eta\right| \le M_1^{1-t}M_2^t \tag{14.9}$$

holds for all simple vectors x on Ω and y on N subject to the restriction $\|x\|_{P,r}=1$, $\|y\|_{Q',s'}=1$. Choose two such vectors $x=(x_1,x_2,\ldots,x_n)$, $y=(y_1,y_2,\ldots,y_m)$ and write $x_k=|x_k|e^{iu_k}$ and $y_k=|y_k|e^{iv_k}$, where u_k and v_k are real-valued simple measurable functions on Ω_k and N_k respectively. Let

$$\alpha_k(z)=\frac{1-z}{p_{1k}}+\frac{z}{p_{2k}}\quad(1\le k\le n),\quad \beta_k'(z)=\frac{1-z}{q_{1k}'}+\frac{z}{q_{2k}'}\quad(1\le k\le m),$$

$$\gamma(z)=\frac{1-z}{r_1}+\frac{z}{r_2},\text{ and }\ \delta'(z)=\frac{1-z}{s_1'}+\frac{z}{s_2'}.$$

Note that

$$\alpha_k(t)=1/p_k,\ \beta_k'(t)=1/q_k',\ \gamma(t)=1/r,\text{ and }\ \delta'(t)=1/s'.$$

For $1\le k\le n$ define

$$X_k(z)=\begin{cases}\|x_k\|_{p_k}^{r\gamma(z)-p_k\alpha_k(z)}\ |x_k|^{p_k\alpha_k(z)}\ e^{iu_k} & \text{if } p_k\ne\infty\\[2mm] \|x_k\|_\infty^{r\gamma(z)-1}\ x_k & \text{if } p_k=\infty\,;\end{cases}$$

for $1\le k\le m$ define

$$Y_k(z)=\begin{cases}\|y_k\|_{q_k'}^{s'\delta'(z)-q_k'\beta_k'(z)}\ |y_k|^{q_k'\beta_k'(z)}\ e^{iv_k} & \text{if } q_k\ne 1\\[2mm] \|y_k\|_\infty^{s'\delta'(z)-1}\ y_k & \text{if } q_k=1,\end{cases}$$

with the understanding that $r\gamma(z)=1$ if $r=\infty$ and $s'\delta'(z)=1$ if $s'=\infty$, and let

$$X(z)=(X_1(z),X_2(z),\ldots,X_n(z)),$$
$$Y(z)=(Y_1(z),Y_2(z),\ldots,Y_m(z)).$$

When $z=t$, $X(t)=x$ and $Y(t)=y$. Finally, define

$$\Phi(z)=\int TX(z)Y(z)d\eta=\sum_{k=1}^{m}\left(\int_{N_k}(TX(z))_k\,Y_k(z)dv_k\right)\cdot\eta_k.$$

Because x and y are simple vectors the components $X_k(z)$ and $Y_k(z)$ are linear combinations of characteristic functions whose coefficients are entire functions bounded in the strip $0\le\operatorname{Re}z\le 1$. It is easy to verify that $X(z)$ can be written as a finite linear combination of vectors, each having only one nonzero component-that being a fixed characteristic function independent of z-with coefficients entire functions bounded in the strip $0\le\operatorname{Re}z\le 1$. Consequently $\Phi(z)$ is an entire function bounded in the strip $0\le\operatorname{Re}z\le 1$.

When $\operatorname{Re}z=1$,

$$\|Y(1+i\tau)\|_{Q_2',s_2'}^{s_2'}=\sum_{k=1}^{m}\left\|\|y_k\|_{q_k'}^{s'/s_2'-q_k'/q_{2k}'}\ |y_k|^{q_k'/q_{2k}'}\right\|_{q_{2k}'}^{s_2'}\eta_k$$

$$=\sum_{k=1}^{m}\|y_k\|_{q_k'}^{s'}\cdot\eta_k=\|y\|_{Q',s'}^{s'}=1\ (-\infty<\tau<\infty);$$

and when $\operatorname{Re} z = 0$,

$$\|Y(i\tau)\|_{Q'_1,s'_1}^{s'_1} = \sum_{k=1}^m \left\| \|y_k\|_{q'_k}^{s'/s'_1 - q'_k/q'_{1k}} |y_k|^{q'_k/q'_{1k}} \right\|_{q'_{1k}}^{s'_1} \cdot \eta_k$$

$$= \sum_{k=1}^m \|y_k\|_{q'_k}^{s'_1} \cdot \eta_k = \|y\|_{Q',s'}^{s'} = 1 \ (-\infty < \tau < \infty).$$

Similar computations show that for all real τ,

$$\|X(1+i\tau)\|_{P_2,r_2} = 1 \text{ and } \|X(i\tau)\|_{P_1,r_1} = 1.$$

The calculations combine to imply that

$$|\Phi(i\tau)| = \left| \int TX(i\tau)Y(i\tau)d\eta \right| \leq \|TX(i\tau)\|_{Q_1,s_1} \ \|Y(i\tau)\|_{Q'_1 s_1}$$

$$\leq M_1 \|X(i\tau)\|_{P_1,r_1} \ \|Y(i\tau)\|_{Q'_1,s'_1} = M_1.$$

Also

$$|\Phi(1+i\tau)| \leq M_2 \|X(1+i\tau)\|_{P_2,r_2} \ \|Y(1+i\tau)\|_{Q'_2,s'_2} = M_2.$$

Thus, by the Phragmen-Lindelöf principle, we get

$$|\Phi(t)| \leq M_1^{1-t} M_2^t.$$

Since this is just (14.9), (14.8) is established. The last assertion of the theorem is clear since the simple vectors are dense in $L^{P,r}(\lambda)$ when $P < \infty$. The proof is now complete. $\square$

§15. L^p Inequalities

Theorem 15.1. *Let (Ω,μ) be a σ-finite measure space. Choose functions $x_1,x_2,\ldots,x_n$ in $L^p(\mu)$ and nonnegative numbers $\xi_1,\xi_2,\ldots,\xi_n$ such that $\Sigma_{j=1}^n \xi_j = 1$. Put $\gamma = \max_{1 \leq j \leq n} (1-\xi_j)$. Then the following inequalities hold:*

$$\sum_{j,k=1}^n \xi_j\xi_k \|x_j-x_k\|_p^{2\alpha} \leq 2\gamma^{2-2\alpha} \sum_{j=1}^n \xi_j \|x_j\|_p^{2\alpha} \ (1 \leq 2\alpha \leq p, \ 1 \leq p \leq 2); \quad (15.1)$$

$$\sum_{j,k=1}^n \xi_j\xi_k \|x_j-x_k\|_p^{2\alpha} \leq 2\gamma^{2-2\alpha} \sum_{j=1}^n \xi_j \|x_j\|_p^{2\alpha} \ (1 \leq 2\alpha \leq p', \ 2 \leq p < \infty); \quad (15.2)$$

$$\gamma^{\beta-2} \sum_{j,k=1}^n \xi_j\xi_k \|x_j-x_k\|_p^{\beta} \geq 2\sum_{j=1}^n \xi_j \left\| x_j - \sum_{k=1}^n \xi_k x_k \right\|_p^{\beta} \ (p' \leq \beta, 1 < p \leq 2);$$

$$\quad (15.3)$$

$$\gamma^{\beta-2} \sum_{j,k=1}^n \xi_j\xi_k \|x_j-x_k\|_p^{\beta} \geq 2\sum_{j=1}^n \xi_j \left\| x_j - \sum_{k=1}^n \xi_k x_k \right\|_p^{\beta} \ (p \leq \beta, 2 \leq p < \infty).$$

$$\quad (15.4)$$

Remark. For subsequent applications it is important to note that each of these inequalities persists if γ is replaced by 1.

Proof. We first establish (15.1) and (15.2). Let $L^{P,r}(\xi)$ denote the space consisting of vectors $x = (x_1,x_2,\ldots,x_n)$ with $x_k \in L^{P_k}(\mu)$ and norm

$$\|x\|_{P,r} = \left(\sum_{k=1}^{n} \|x_k\|_{P_k}^r \, \xi_k \right)^{1/r}.$$

In addition let $L^{Q,s}(\xi \times \xi)$ denote the vectors $y = (y_{jk})_{j,k=1}^{n}$ with $y_{jk} \in L^{q_{jk}}(\mu)$ and norm

$$\|y\|_{Q,s} = \left(\sum_{j,k=1}^{n} \xi_j \xi_k \, \|y_{jk}\|_{q_{jk}}^s \right)^{1/s}.$$

It is no restriction to suppose that the ξ_k are positive.

Define a map T from the simple measurable vectors in $L^{P,r}(\xi)$ to the measurable vectors in $L^{Q,s}(\xi \times \xi)$ by

$$T : (x_1, x_2, \ldots, x_n) \rightarrow (x_j - x_k)_{j,k=1}^{n}.$$

In order to simplify notation, let a multi-index of equal components be denoted by that component. Put

$$P_1 = 2, \; r_1 = 2, \; Q_1 = 2, \; s_1 = 2, \; M_1 = \sqrt{2}$$

$$P_2 = 1, \; r_2 = 1, \; Q_2 = 1, \; s_2 = 1, \; M_2 = 2\gamma.$$

The inequalities corresponding to (14.6) and (14.7) are

$$\|Tx\|_{2,2}^2 = \sum_{j,k=1}^{n} \xi_j \xi_k \, \|x_j - x_k\|_2^2 = 2 \sum_{j=1}^{n} \xi_j \, \|x_j\|_2^2 - 2 \left\| \sum_{j=1}^{n} \xi_j x_j \right\|_2^2 \qquad (15.5)$$

$$\leq 2 \sum_{j=1}^{n} \xi_j \, \|x_j\|_2^2 = 2 \|x\|_{2,2}^2,$$

and

$$\|Tx\|_{1,1} = \sum_{j,k=1}^{n} \xi_j \xi_k \, \|x_j - x_k\|_1 \qquad (15.6)$$

$$\leq \sum_{j,k=1}^{n} \xi_j \xi_k (\|x_j\|_1 + \|x_k\|_1) - 2 \sum_{j=1}^{n} \xi_j^2 \, \|x_j\|_1$$

$$= 2 \sum_{j=1}^{n} \xi_j (1 - \xi_j) \, \|x_j\|_1 \leq 2\gamma \sum_{j=1}^{n} \xi_j \, \|x_j\|_1 = 2\gamma \|x\|_{1,1}.$$

For $0 < t < 1$ put $1/p = t + (1-t)/2 = (1+t)/2$, whence $P = p$, $r = p$, $Q = p$ and $s = p$. It follows from the interpolation theorem and (14.8) that

$$\|Tx\|_{p,p} \leq 2^{1/p} \gamma^{-1+2/p} \|x\|_{p,p} \qquad (x \in L^{P,r}(\xi)). \qquad (15.7)$$

This is (15.1) with $\alpha = p/2$. The general case follows from a second application of the interpolation theorem, this time with endpoint conditions (15.7) and the obvious analogue of (15.6),

$$\|Tx\|_{p,1} \leq 2\gamma \|x\|_{p,1} \qquad (x \in L^{P,1}(\xi)). \qquad (15.8)$$

valid for $1 \leq p \leq \infty$. In this case the indices and constants are

$$P_1 = p, \; r_1 = p, \; Q_1 = p, \; s_1 = p, \; M_1 = 2^{1/p} \, \gamma^{-1+2/p}$$

$$P_2 = p, \; r_2 = 1, \; Q_2 = p, \; s_2 = 1, \; M_2 = 2\gamma,$$

$P = Q = p$, and $r = s = 2\alpha$ where $1/2\alpha = t + (1-t)/p$, $0 < t < 1$. By (14.8), .

$$\|Tx\|_{p,2\alpha} \leq 2^{1/2\alpha} \gamma^{-1+1/\alpha} \|x\|_{p,2\alpha} \qquad (x \in L^{p,r}(\xi)).$$

This proves (15.1). The inequality (15.2) follows in much the same way. Define T as before and interpolate between (15.5) and (15.8) with $p = \infty$ to get

$$\|Tx\|_{p,p'} \leq 2^{1/p'} \gamma^{-1+2/p'} \|x\|_{p,p'} \qquad (15.9)$$

for $2 \leq p < \infty$. This is (15.2) with $2\alpha = p'$. The general case, $1 \leq 2\alpha \leq p'$, results from a second interpolation with (15.9) and (15.8) as endpoint conditions.

The remaining inequalities require a more elaborate setting. For $1 \leq p, r \leq \infty$, let $E^{p,r} = L^{p,r}(\xi_1\xi_2,\ldots,\xi_1\xi_n,\xi_2\xi_3,\ldots,\xi_{n-1}\xi_n)$ denote the space of all $(n-1)$-tuples $x = (x^1, x^2,\ldots,x^{n-1})$ in which each x^j is an $(n-j)$-tuple of elements of $L^p(\mu)$. On this space introduce the norm

$$\|x\|_{p,r} = \left(\sum_{j=1}^{n-1} \sum_{k>j} \xi_j \xi_k \|x_{k-j}^j\|_p^r \right)^{1/r}$$

and

$$\|x\|_{p,\infty} = \max\{\|x_{k-j}^j\|_p : 1 \leq j \leq n-1, k > j\}.$$

Consider the linear operator T defined from $E^{p,r}$ to $L^{p,r}(\xi)$ by

$$(Tx)_j = \sum_{i=1}^{n-j} \xi_{i+j} x_i^j - \sum_{k=1}^{j-1} \xi_k x_{j-k}^k, \quad 1 \leq j \leq n,$$

it being understood that a summation is zero if the upper index is zero.

In order to apply the interpolation theorem we need to verify that

$$\|Tx\|_{2,2} \leq \|x\|_{2,2} \qquad (15.10)$$

and

$$\|Tx\|_{q,\infty} \leq \gamma \|x\|_{q,\infty} \qquad (1 \leq q \leq \infty) \qquad (15.11)$$

for all simple measurable vectors x.

The second inequality comes from the estimate

$$\|Tx\|_{q,\infty} = \max_{1 \leq j \leq n} \left\| \sum_{i=1}^{n-j} \xi_{i+j} x_i^j - \sum_{k=1}^{j-1} \xi_k x_{j-k}^k \right\|_q$$

$$\leq \max_{1 \leq j \leq n} \left\{ \sum_{i=1}^{n-j} \xi_{i+j} \|x_i^j\|_q + \sum_{k=1}^{j-1} \xi_k \|x_{j-k}^k\|_q \right\}$$

$$\leq \max\{(1-\xi_j) \|x\|_{q,\infty} : 1 \leq j \leq n\} \leq \gamma \|x\|_{q,\infty}.$$

The demonstration of (15.10) is rather tedious. From the definition of T and the norm $\|\cdot\|_{2,2}$, we have

$$\|Tx\|_{2,2}^2 = \sum_{j=1}^{n} \xi_j \left\| \sum_{i=1}^{n-j} \xi_{i+j} x_i^j - \sum_{k=1}^{j-1} \xi_k x_{j-k}^k \right\|_2^2$$

$$= \sum_{j=1}^{n} \xi_j \left\{ \sum_{i,\ell=1}^{n-j} \xi_{i+j} \xi_{+j} (x_i^j, x_\ell^j) + \sum_{k=1}^{j-1} \sum_{\ell=1}^{n-j} \xi_k \xi_{\ell+j} (x_{j-k}^k, x_\ell^j) \right.$$

$$\left. - \sum_{m=1}^{j-1} \sum_{i=1}^{n-j} \xi_{i+j} \xi_m (x_i^j, x_{j-m}^m) + \sum_{k,m=1}^{j-1} \xi_k \xi_m (x_{j-k}^k, x_{j-m}^m) \right\}$$

$$= \sum_{j=1}^n \sum_{i,k=1}^{n-j} \xi_j \xi_{i+j} \xi_{k+j} (x_i^j, x_k^j) + \sum_{j=1}^n \sum_{i,k=1}^{j-1} \xi_j \xi_i \xi_k (x_{j-k}^k, x_{j-i}^i)$$

$$- \sum_{j=1}^n \sum_{k=1}^{j-1} \sum_{i=1}^{n-j} \xi_j \xi_k \xi_{i+j} \{(x_{j-k}^k, x_i^i) + (x_i^i, x_{j-k}^k)\}.$$

Of course, $(\cdot, \cdot)$ denotes the inner product in $L^2(\mu)$. After some rearrangement this sum takes the form

$$\|Tx\|_{2,2}^2 = \sum_{1 \le i < j < k \le n} \xi_i \xi_j \xi_k \left\{ (x_{j-i}^i, x_{k-i}^i) + (x_{k-i}^i, x_{j-i}^i) + (x_{k-i}^i, x_{k-j}^j) \right.$$

$$+ (x_{k-j}^j, x_{k-i}^i) - (x_{j-i}^i, x_{k-j}^j) - (x_{k-j}^j, x_{j-i}^i) \Big\}$$

$$+ \sum_{\substack{i=1 \\ j>i}}^n \left\{ \xi_i \xi_j \xi_j (x_{j-i}^i, x_{j-i}^i) + \xi_i \xi_i \xi_j (x_{j-i}^i, x_{j-i}^i) \right\}.$$

This last sum is equal to

$$\sum_{\substack{i=1 \\ j>i}}^n \left(1 - \sum_{k \ne i,j} \xi_k \right) \xi_i \xi_j (x_{j-i}^i, x_{j-i}^i) = \|x\|_{2,2}^2 - \sum_{\substack{i=1 \\ j>i}}^{n-1} \sum_{k \ne i,j} \xi_i \xi_j \xi_k (x_{j-i}^i, x_{j-i}^i).$$

Hence,

$$\|Tx\|_{2,2}^2 = \|x\|_{2,2}^2 - \sum_{1 \le i < j < k \le n} \xi_i \xi_j \xi_k \left\{ -(x_{j-i}^i, x_{k-i}^i) - (x_{k-i}^i, x_{j-i}^i) \right.$$

$$- (x_{k-i}^i, x_{k-j}^j) - (x_{k-j}^j, x_{k-i}^i) + (x_{j-i}^i, x_{k-j}^j) + (x_{k-j}^j, x_{j-i}^i)$$

$$+ (x_{j-i}^i, x_{j-i}^i) + (x_{k-j}^j, x_{k-j}^j) + (x_{k-i}^i, x_{k-i}^i) \Big\}$$

$$= \|x\|_{2,2}^2 - \sum_{1 \le i < j < k \le n} \xi_i \xi_j \xi_k \|x_{j-i}^i - x_{k-i}^i + x_{k-j}^j\|_2^2 \le \|x\|_{2,2}^2.$$

In order to apply the interpolation theorem to the operator T, suppose $1 < p \le 2$, $p' \le \beta < \infty$ and $1 - t = 2/\beta$. From the inequalities

$$(1-t)/2 + t = (1+t)/2 = 1 - (1-t)/2 = 1 - 1/\beta \ge 1 - 1/p'$$

$$= 1/p = (1-t)/p + t/p \ge (1-t)/2 + t/p$$

it follows that there is a number r, $1 \le r \le p$, such that $1/p = (1-t)/2 + t/r$.

Now apply Theorem 14.1 to (15.10) and (15.11) with $q = r$ to obtain

$$\|Tx\|_{p,\beta} \le \gamma^{1-2/\beta} \|x\|_{p,\beta} \qquad (1 < p \le 2, \; p' \le \beta). \tag{15.12}$$

A similar argument shows that (15.12) also holds for $2 \le p < \infty$ and $p \le \beta < \infty$.

The inequalities (15.3) and (15.4) are a direct consequence of (15.12). Choose simple measurable functions $x_1, x_2, \ldots, x_n$ in $L^p(\mu)$ and put $x = (x_1 - x_2, x_1 - x_3, \ldots, x_1 - x_n, x_2 - x_3, \ldots, x_{n-1} - x_n)$ so that $x_i^j = x_j - x_{i+j}$. Having made these identifications, we proceed to calculate the norms in (15.12):

$$\|Tx\|_{p,\beta}^\beta = \sum_{j=1}^n \xi_j \left\| \sum_{i=1}^{n-j} \xi_{i+j} (x_j - x_{i+j}) - \sum_{k=1}^{j-1} \xi_k (x_k - x_j) \right\|_p^\beta$$

$$= \sum_{j=1}^n \xi_j \left\| \sum_{i=1}^n \xi_i (x_j - x_i) \right\|_p^\beta = \sum_{j=1}^n \xi_j \left\| x_j - \sum_{i=1}^n \xi_i x_i \right\|_p^\beta.$$

$$\|x\|_{p,\beta}^{\beta} = \sum_{i=1}^{n-1} \sum_{j>i} \xi_i \xi_j \|x_i - x_j\|_p^{\beta} = \frac{1}{2} \sum_{i,j=1}^{n} \xi_i \xi_j \|x_i - x_j\|_p^{\beta}. \quad \square$$

The inequalities just established have obvious implications concerning the sums of powers of distances between the points of finite sets in L^p spaces.

Corollary 15.2. *If* $x_1, x_2, \ldots, x_n$ *are unit vectors in* L^p *such that* $\sum_{i=1}^n x_i = 0$, *then*

$$\sum_{j,k=1}^{n} \|x_j - x_k\|_p^p \geq 2n^p(n-1)^{2-p} \quad \text{for} \quad 2 \leq p < \infty$$

and

$$\sum_{j,k=1}^{n} \|x_j - x_k\|_p^{p'} \geq 2n^{p'}(n-1)^{2-p'} \quad \text{for} \quad 1 < p \leq 2.$$

Proof. Apply the inequalities (15.3) and (15.4) with $\xi_j = 1/n$ and β equal first p' and then p. $\square$

Such inequalities have been studied with better results by Chakerian and Klamkin [15] when $p = 2$. They show that

$$\sum_{j,k=1}^{n} \|x_j - x_k\|_2^2 > 4(n-1)$$

for any set of unit vectors in Hilbert space provided only that the origin lies in their convex hull.

Corollary 15.3. *If* $x, x_1, x_2, \ldots, x_n$ *are points in* L^p *such that* $\|x_j - x_k\|_p \geq 2$ $(1 \leq j, k \leq n)$, *then*

$$\sum_{j=1}^{n} \|x_j - x\|_p^p \geq 2^{p-1}(n-1)^{p-1} n^{2-p} \quad \text{for} \quad 1 \leq p \leq 2,$$

and

$$\sum_{j=1}^{n} \|x_j - x\|_p^{p'} \geq 2^{p'-1}(n-1)^{p'-1} n^{2-p'} \quad \text{for} \quad 2 \leq p < \infty.$$

Proof. Take $\xi_j = \frac{1}{n}$ and $2\alpha = p$ and p' respectively in (15.1) and (15.2). $\square$

One consequence is that the radius r of a sphere in L^p which contains n points each at least a distance of 2 from all the others must satisfy

$$r > 2^{1/p'}(1 - 1/n)^{1/p'} \text{ if } 1 < p \leq 2 \text{ and } r \geq 2^{1/p}(1 - 1/n)^{1/p} \text{ if } 2 \leq p < \infty.$$

This observation is due to H. F. Blichfeldt [6] in the case $p = 2$.

Other important L^p inequalities may be derived with the aid of Theorem 15.1.

Theorem 15.4 *If* $x_1, x_2 \in L^p(\mu)$ *and* a *is a complex number, then*

$$(\|x_1 + ax_2\|_p^{p'} + \|\bar{a}x_1 - x_2\|_p^{p'})^{1/p'} \tag{15.13}$$

$$\leq [\max(1, |a|)]^{-1 + 2/p}(1 + |a|^2)^{1 - 1/p}(\|x_1\|_p^p + \|x_2\|_p^p)^{1/p}$$

for $1 < p \leq 2$,

and

$$(\|x_1 + ax_2\|_p^p + \|\bar{a}x_1 - x_2\|_p^p)^{1/p} \tag{15.14}$$

$$\leq [\max(1,|a|)]^{1-2/p}(1+|a|^2)^{1/p}(\|x_1\|_p^{p'} + \|x_2\|_p^{p'})^{1/p'}$$

for $2 \leq p < \infty$.

Proof. Define a linear operator T on pairs of simple measurable functions by

$$T : (x_1, x_2) \to (x_1 + ax_2, \bar{a}x_1 - x_2),$$

and take $\lambda_k = \eta_k = 1$ $(k = 1,2)$ in (14.2). We have the obvious inequalities:

$$\|T(x_1,x_2)\|_{(2,2),2} = (1+|a|^2)^{1/2} \|(x_1,x_2)\|_{(2,2),2}. \tag{15.15}$$

$$\|T(x_1,x_2)\|_{(1,1),\infty} \leq \max(1,|a|) \|(x_1,x_2)\|_{(1,1),1}. \tag{15.16}$$

$$\|T(x_1,x_2)\|_{(\infty,\infty),\infty} \leq \max(1,|a|) \|(x_1,x_2)\|_{(\infty,\infty),1}. \tag{15.17}$$

Interpolation with (15.15) and (15.16) as endpoint conditions gives (15.13), and (15.14) follows from applying Theorem 14.1 to (15.15) and (15.17). $\square$

When $a = 1$, (15.13) and (15.14) reduce to inequalities of Clarkson. For references and related ideas see [22, p.473]. Our second application is due to Harris [36].

Theorem 15.5. *For* $1 \leq p \leq \infty$, *put*

$$K_p(n) = [4^n \binom{2n}{n}^{-1}]^{|p-2|/p} \qquad (n = 1,2,\ldots),$$

it being understood that the exponent is 1 when $p = \infty$. *Then if* $\lambda = \exp(i\pi/n)$ *and* $x, y \in L^p(\mu)$,

$$\sum_{k=1}^{2n} \|x + \lambda^k y\|_p^{2n} \leq n \binom{2n}{n} K_p(n) (\|x\|_p^{2n} + \|y\|_p^{2n}). \tag{15.18}$$

Proof. It is evident from the inequality $2n(1+t)^{2n} \leq n4^n(1+t^{2n})$ $(0 \leq t \leq 1)$ that

$$\sum_{k=1}^{2n} \|x + \lambda^k y\|_p^{2n} \leq n4^n(\|x\|_p^{2n} + \|y\|_p^{2n}) \tag{15.19}$$

when $p = 1$ and $p = +\infty$. Further,

$$\sum_{k=1}^{2n} \|x + \lambda^k y\|_2^{2n} \leq n \binom{2n}{n} (\|x\|_2^{2n} + \|y\|_2^{2n}). \tag{15.20}$$

But this is not so evident. First assume $\|x\|_2 = 1$ and $\|y\|_2 = r$. By the binomial theorem,

$$\|x + \lambda^k y\|_2^{2n} = \sum_{\ell=0}^{n} \binom{n}{\ell} (1+r^2)^{n-\ell} \, 2^{\ell} \, (\mathrm{Re} \, \bar{\lambda}^k(x,y))^{\ell}.$$

Put $\sigma = (x,y)/|(x,y)|$ if $(x,y) \neq 0$ and $\sigma = 1$ if $(x,y) = 0$.

Then $|\mathrm{Re} \, \bar{\lambda}^k(x,y)| \leq r |\mathrm{Re} \, \bar{\lambda}^k \sigma|$ and

$$\sum_{k=1}^{2n} (\mathrm{Re} \, \bar{\lambda}^k \sigma)^{\ell} = \sum_{k=1}^{n} (\mathrm{Re} \, \bar{\lambda}^k \sigma)^{\ell} + \sum_{k=1}^{n} (-\mathrm{Re} \, \bar{\lambda}^k \sigma)^{\ell} = 0$$

when ℓ is odd. Hence

$$\sum_{k=1}^{2n} \|x + \bar{\lambda}^k y\|_2^{2n} \leq \sum_{k=1}^{2n} \sum_{\ell=0}^{n} \binom{n}{\ell}(1+r^2)^{n-\ell}\, 2^\ell\, r^\ell\, (\text{Re}\,\bar{\lambda}^k\sigma)^\ell \qquad (15.21)$$

$$= \sum_{k=1}^{2n} |1 + r\bar{\lambda}^k\sigma|^{2n}.$$

Again, by the binomial theorem,

$$|1 + r\bar{\lambda}^k\sigma|^{2n} = \sum_{\ell=0}^{n} \sum_{m=0}^{n} \binom{n}{\ell}\binom{n}{m} r^{\ell+m}\,\sigma^{\ell-m}\,\bar{\lambda}^{k(\ell-m)}.$$

Since

$$\sum_{k=1}^{2n} \lambda^{kj} = \begin{cases} 0 & \text{if } j \neq 0 \\ 2n & \text{if } j = 0 \end{cases}$$

for $-n \leq j \leq n$, we may sum both sides of the last expansion to get

$$\sum_{k=1}^{2n} |1 + r\bar{\lambda}^k\sigma|^{2n} = 2n\sum_{m=0}^{n} \binom{n}{m}^2 r^{2m} \leq n\sum_{m=0}^{n} \binom{n}{m}^2 (1+r^{2n}), \qquad (15.22)$$

the last inequality resulting from the symmetry of the binomial coefficients and the convexity inequality

$$r^{2(n-m)} + r^{2m} \leq 1 + r^{2n}.$$

The inequality (15.20) now follows from (15.21), (15.22) and the identity

$$\sum_{m=0}^{n} \binom{n}{m}^2 = \binom{2n}{n}.$$

In the context of Theorem 14.1 we define an operator T on pairs of simple vectors by

$$T : (x,y) \to (x + \lambda y, x + \lambda^2 y, \dots, x + \lambda^{2n} y)$$

with $\lambda_1 = \lambda_2 = 1$ and $\eta_1 = \eta_2 = \dots = \eta_{2n} = 1$. The inequalities (15.19) and (15.20) take the form

$$\|T(x,y)\|_{(p,p,\dots,p),2n} \leq 2n^{1/2n} \|(x,y)\|_{(p,p),2n} \quad (p=1, +\infty) \qquad (15.23)$$

and

$$\|T(x,y)\|_{(2,2,\dots,2),2n} \leq n^{1/2n} \binom{2n}{n}^{1/2n} \|(x,y)\|_{(2,2),2n}. \qquad (15.24)$$

Two interpolations between (15.24) and (15.23), with p alternately 1 and ∞, yield (15.18). $\square$

From (15.18) it is easy to obtain the inequality

$$\frac{1}{\pi}\int_0^{2\pi} \|x + e^{i\theta}y\|_p^{2n}\,d\theta \leq \binom{2n}{n}K_p(n)\,(\|x\|_p^{2n} + \|y\|_p^{2n}) \qquad (15.25)$$

for $x,y \in L^p(\mu)$ $(1 \leq p \leq \infty)$.

It should be pointed out that both (15.25) and (15.18) are sharp when $p = 2$, for equality holds if L^2 is the complex plane and $x = y = 1$.

These inequalities are used in [36] to estimate the norms of homogeneous polynomials associated with the nth-order Frechet derivatives of holomorphic functions on the unit ball in L^p spaces.

§16. A Packing Problem in L^p

The packing problem we consider in this section is a special case of the following: How many balls of radius r can be packed in the unit ball of a Banach space? Mainly we are interested in showing how inequalities (15.1) and (15.2) provided an elegant solution to this problem in L^p spaces. Here L^p denotes the Banach space $L^p(\Omega, \mu)$ $(1 \leq p \leq \infty)$ over some σ-finite measure space. This problem was studied in ℓ^p by Burlak, Rankin and Robertson [14] and our results are patterned after theirs.

Definition 16.1. *Let U_p denote the unit ball in L^p. We shall say that a collection of balls $\{B(x_j, r)\}$,*

$$B(x_j, r) = \{y \in L^p : \|y - x_j\| \leq r\},$$

is packed in U_p provided that

(a) *$B(x_j, r) \subset U_p$ for each index j*

and

(b) *the interiors of any two of the balls are disjoint,*

$$\text{int } B(x_i, r) \cap \text{int } (B(x_j, r)) = \emptyset \qquad (i \neq j).$$

The conditions for packing translate into the inequalities

(a') $\|x_j\|_p \leq 1 - r$ for all j,

and

(b') $\|x_j - x_k\|_p \geq 2r$ for $j \neq k$.

If L^p is infinite-dimensional there must be a cut $\lambda(U_p)$ $(0 < \lambda(U_p) \leq 1)$ such that infinitely many balls of radius r can be packed in U_p when $0 \leq r \leq \lambda(U_p)$ and only finitely many when $\lambda(U_p) < r \leq 1$. In fact we can calculate, or rather guess, a lower bound λ_p for $\lambda(U_p)$ by means of a simple example.

To do this construct a sequence $\{x_j\}_{j=1}^\infty$ in L^p $(1 \leq p < \infty)$ such that

$$\|x_j\|_p = 1 - r \ (j \geq 1) \text{ and } \|x_j - x_k\|_p = (1 - r) \, 2^{1/p} \ (j \neq k).$$

This may be done by choosing a sequence $\{E_j\}_{j=1}^\infty$ of measurable and pairwise disjoint sets in Ω of positive finite measure and letting $y_j = \mu(E_j)^{-1} \chi_{E_j}, j = 1, 2, \ldots$ (see Theorem 7.5). Fix r, $0 < r \leq 1$ and put $x_j = (1 - r)y_j$ for $j = 1, 2, \ldots$. The centers of the family $\{B(x_j, r)\}$ satisfy (a') and

$$\|x_j - x_k\|_p = (1 - r)2^{1/p} \qquad (j \neq k).$$

If (b') is to hold we must have $(1 - r)2^{1/p} \geq 2r$ or

$$r \le \frac{1}{1+2^{1-1/p}} = \lambda_{p'}. \tag{16.1}$$

Thus infinite packing is possible in L^p ($1 \le p < \infty$) whenever r satisfies (16.1).

Theorem 16.2. *Suppose $1 \le p \le 2$ and L^p is infinite-dimensional. Then $\lambda(U_p) = \lambda_p$, that is, an infinite number of balls of radius r can be packed in U_p if and only if r satisfies (16.1). If $\lambda_p < r \le 1$, then at most a finite number of spheres of radius r can be packed in U_p and that number does not exceed*

$$L_{p'}(r) = \left[1 - \frac{1}{2}\left(\frac{1-r}{r}\right)^{p'}\right]^{-1} \quad for \quad 1 < p \le 2,$$

$$L_1(r) = 1 \quad when \quad p = 1.$$

Proof. Our opening remarks show that infinite packing is possible when $r \le \lambda_p$. What remains is to show that infinite packing is possible only if $r \le \lambda_p$.

Suppose $0 < r \le 1$ and the balls $\{B(x_j, r)\}_{j=1}^n$ are packed in U_p. Take $2\alpha = p$ and $\xi_j = \frac{1}{n}$ in (15.1) so that $\gamma = 1 - \frac{1}{n}$ and the inequality becomes

$$\sum_{j,k=1}^n \|x_j - x_k\|_p^p \le 2 \, n^{p-1} \, (n-1)^{2-p} \sum_{j=1}^n \|x_j\|_p^p.$$

The packing inequalities (a') and (b') combine with this to give

$$n(n-1) \, 2^p \, r^p \le 2 \, n^{p-1}(n-1)^{2-p} \, n(1-r)^p \tag{16.2}$$

or

$$r \le [1 + (1 - 1/n)^{1-1/p} \, 2^{1-1/p}]^{-1}.$$

The inequality $r \le \lambda_p$ results from letting $n \to \infty$, and solving (16.2) for n gives

$$n \le \left[1 - \frac{1}{2}\left(\frac{1-r}{r}\right)^{p'}\right]^{-1} \quad for \quad 1 < p \le 2.$$

The situation for $p = 1$ is evident: infinite packing occurs when $0 < r \le \frac{1}{2}$, but only one sphere of radius r can be packed in U_1 when $\frac{1}{2} < r \le 1$. $\square$

In the range $2 < p \le \infty$ there is a more subtle dependence on the underlying measure space.

Theorem 16.3. *Suppose $2 < p < \infty$ and Ω contains a set E of positive and finite measure each of whose measurable subsets F_1 contains a measurable subset F_2 such that $2\mu(F_2) = \mu(F_1)$. Then an infinite number of balls of radius r can be packed in U_p if and only if*

$$0 < r \le \frac{1}{1+2^{1/p}} = \lambda_{p'}. \tag{16.3}$$

Thus $\lambda(U_p) = \lambda_{p'}$. If $\lambda_{p'} < r \le 1$, then only a finite number of balls of radius r can be packed in U_p and that number does not exceed

$$L_p(r) = \left[1 - \frac{1}{2}\left(\frac{1-r}{r}\right)^p\right]^{-1}.$$

Proof. Suppose that the finite collection $\{B(x_j,r)\}_{j=1}^n$ is packed in U_p. Take $2\alpha = p'$ and $\xi_j = 1/n$ in (15.2) so that $\gamma = 1 - 1/n$ and the inequality becomes

$$\sum_{j,k=1}^n \|x_j - x_k\|_p^{p'} \le 2n^{p'/p} (n-1)^{1-p'/p} \sum_{j=1}^n \|x_j\|_p^{p'}.$$

The packing inequalities (a') and (b') combine with this to give

$$n(n-1) \, 2^{p'} \, r^{p'} \le 2n^{p'/p} (n-1)^{1-p'/p} \, n(1-r)^{p'}, \tag{16.4}$$

or

$$r < [1 + (1-1/n)^{1/p} \, 2^{1/p}]^{-1}.$$

The inequality (16.3) results from letting $n \to \infty$ and solving (16.4) for n; we get

$$n \le \left[1 - \frac{1}{2}\left(\frac{1-r}{r}\right)^p\right]^{-1},$$

which is the required estimate when $\lambda_{p'} < r \le 1$. It remains to show that infinite packing is always possible for $0 < r \le \lambda_{p'}$.

As a consequence of our assumption on E, these exists, for each positive integer n, pairwise disjoint subsets $\{E_{n,j}\}_{j=1}^{2n}$ such that $\bigcup_{j=1}^n E_{n,j} = E$ and

$$E_{n-1,k} \supset E_{n,2k-1} \cup E_{n,2k} \text{ and } \mu(E_{n,2k-1}) = \mu(E_{n,2k}) = \tfrac{1}{2}\mu(E_{n-1,k})$$

for $k = 1, 2, \ldots, 2^{n-1}$. Using these sets we define "Rademacher" functions $\{y_n\}_{n=1}^\infty$ by the formula

$$y_n = \mu(E)^{-1/p} \sum_{k=1}^{2n} (-1)^{k+1} \chi_{E_{n,k}}.$$

The members of this sequence satisfy

$$\|y_n\|_p = 1 \text{ and } \|y_j - y_k\|_p = 2^{1-1/p} \quad (j \ne k).$$

Now consider the balls $\{B(x_j,r)\}_{j=1}^\infty$ where $x_j = (1-r)y_j (j \ge 1)$. We have $\|x_j\|_p = 1-r$ and $\|x_j - x_k\|_p = (1-r)2^{1-1/p}$. The packing conditions will be satisfied provided that $(1-r)2^{1-1/p} \ge 2r$ or

$$r \le \frac{1}{1+2^{1/p}} = \lambda_{p'}. \quad \square$$

The situations in L^1 and L^∞ are identical.

Theorem 16.4. *Suppose L^∞ is infinite-dimensional. Then infinitely many balls of radius r can be packed in U_∞ provided $0 < r \le 1/2$, and exactly one ball of radius r can be packed in U_∞ if $1/2 < r < 1$.*

Proof. The last assertion is a consequence of the triangular inequality. To show that infinite packing is possible when $0 < r \le 1/2$ choose infinitely many pairwise disjoint subsets E_j of Ω of positive and finite measure, then define $x_j = \tfrac{1}{2}\chi_{E_j}$ $(j = 1, 2, \ldots)$. The infinite family $\{B(x_j, 1/2)\}_{j=1}^\infty$ is clearly packed in U_∞. $\square$

The rather awkward initial assumption in Theorem 16.3 is not frivolous, indeed the absence of Rademacher-like functions in L^p $(2 < p < \infty)$ works a dramatic change in the conclusion when μ consists only of point masses.

Theorem 16.5. *If $2 < p < \infty$· then infinitely many balls can be packed in the unit ball of ℓ^p if and only if $0 < r \leq \lambda_p$. Thus $\lambda(U_p) = \lambda_p$. If $\lambda_p < r \leq \lambda_{p'}$, then any finite number of balls of radius r can be packed in U_p and if $\lambda_{p'} < r \leq 1$, the number of balls of radius r which can be packed in U_p does not exceed*

$$L_p(r) = \left[1 - \frac{1}{2} \left(\frac{1-r}{r} \right)^p \right]^{-1}.$$

Proof. The last assertion is identical with the corresponding conclusion of Theorem 16.3. Hence there are two points to establish:

(1) if infinitely many spheres of radius r can be packed in U_p, then $0 < r \leq \lambda_p$,

and

(2) if $\lambda_p < r \leq \lambda_{p'}$, any finite number of spheres of radius r can be packed in U_p.

To prove (1) suppose infinitely many spheres of radius r are packed in U_p. Let

$$x^j = (x_{j1}, x_{j2}, \ldots) \qquad (j = 1, 2, 3, \ldots)$$

be their centers and let

$$x = (x_1, x_2, \ldots)$$

be a weak limit of the sequence $\{x^j\}$. (We may as well assume the centers themselves converge weakly.) Then $x \in \ell^p$ and $\|x\|_p \leq 1 - r$.

Let $\varepsilon > 0$ and fix a positive integer n. There exists a positive integer N such that

$$\sum_{j > N} |x_{nj}|^p < \varepsilon^p.$$

Then

$$(2r)^p \leq \sum_{j=1}^{\infty} |x_{mj} - x_{nj}|^p = \sum_{j=1}^{N} |x_{mj} - x_{nj}|^p + \sum_{j > N} |x_{mj} - x_{nj}|^p$$

$$\leq \sum_{j=1}^{N} |x_{mj} - x_{nj}|^p + (1 - r + \varepsilon)^p,$$

or

$$(2r)^p - (1 - r + \varepsilon)^p \leq \sum_{j=1}^{N} |x_{mj} - x_{nj}|^p.$$

Since this inequality is independent of m and $x^j \to x$ weakly we may conclude that

$$(2r)^p - (1 - r + \varepsilon)^p \leq \sum_{j=1}^{N} |x_j - x_{nj}|^p.$$

Then letting $N \to \infty$ and $\varepsilon \downarrow 0$ we get

$$(2r)^p - (1 - r)^p \leq \sum_{j=1}^{\infty} |x_j - x_{nj}|^p$$

holding for every n. Again let $\varepsilon > 0$ and choose N so that $\sum_{j > N} |x_j|^p < \varepsilon^p$. From this last inequality,

$$(2r)^p - (1-r)^p \leq \sum_{j=1}^N |x_j - x_{nj}|^p + (\varepsilon + (1-r))^p.$$

Letting $n \to \infty$ and then $\varepsilon \downarrow 0$, this inequality gives

$$(2r)^p - (1-r)^p \leq (1-r)^p,$$
$$(2r)^p \leq 2(1-r)^p,$$
$$2r \leq 2^{1/p}(1-r),$$
$$r \leq \frac{1}{1+2^{1-1/p}} = \lambda_p. \quad \text{This proves (1).}$$

The proof of (2) follows from the observation that the process of constructing Rademacher functions, which we utilized in Theorem 16.3, can be mimicked in ℓ^p to any finite stage through the device of replacing E by the block $\{1,2,\ldots,2^n\}$. $\square$

It is no accident that the packing numbers λ_p of the preceding results all belong to the interval $[1/3, 1/2]$. It is a result of Kottman [50] that such must always be the case for an infinite-dimensional Banach space.

Definition 16.6.   *Let X be a normed linear space with closed unit ball U. We define*

$$\lambda(U) = \sup\{r : \text{infinitely many balls of radius } r \text{ can be packed in } U\},$$

with $\lambda(U) = 0$ in case infinite packing is impossible.

Theorem 16.7.   *If X is an infinite-dimensional Banach space, then*

$$\tfrac{1}{3} \leq \lambda(U) \leq \tfrac{1}{2}.$$

Proof.   The triangular inequality shows that two balls of radius $r > 1/2$ cannot be packed in U, hence $\lambda(U) \leq \tfrac{1}{2}$.

Choose δ, $0 < \delta < 1$, and let S denote the unit sphere $S = \{x \in X : \|x\| = 1\}$. In S define

$$\mathscr{D} = \{D \subset S : x, y \in D \text{ implies } \|x - y\| \geq \delta\}.$$

It is clear by Zorn's lemma that $\mathscr{D}$ has a maximal element D with respect to ordering by inclusion. That D is a δ-net for S (that is, every point of S is at a distance $\leq \delta$ from some point of D) follows from its maximality. Consequently, the closed linear span of D must equal X, for otherwise there would exist a point $y \in S$ such that $\text{dist}(y, D) > \delta$. Hence D cannot be finite. Now consider the infinite collection of balls

$$\{\tfrac{2}{3}y + (\delta - \tfrac{2}{3})U : y \in D\}.$$

The packing conditions (a') and (b') are satisfied. Therefore by letting $\delta \to 1$ we can get an infinite packing by balls of radius as near 1/3 as we please. $\square$

The preceding results suggest a notion of non-compactness. See [48] for properties of ε-separated sets and related concepts.

Definition 16.8.   *A subset of a metric space (X, ρ) is ε-separated ($\varepsilon > 0$) provided no two of its points are closer than ε. And for each bounded subset E of X, define*

$$\delta(E) = \sup\{\varepsilon : E \text{ contains an infinite } \varepsilon\text{-separated subset}\},$$

with $\sup \emptyset = 0$.

There is a striking difference between this definition of non-compactness and the measure of non-compactness γ developed by Kuratowski.

For each bounded subset E of X,

$$\gamma(E) = \inf\{d > 0 : E \text{ can be partitioned into a finite number of sets}$$
$$\text{of diameter} \leq d\}.$$

Properties of $\gamma(E)$ are developed in [19] and [69]. According to a result of R. Nussbaum,

$$\gamma(U) = 2$$

for the unit ball U of any infinite-dimensional normed linear space. After the obvious alterations in the proof of Theorems 16.2–16.4, we have the following result.

Theorem 16.9. *Let U_p denote the unit ball in L^p or ℓ^p. Then*

$$\delta(U_p) = 2^{1/p} \text{ if } 1 \leq p \leq 2,$$

and when the measure space Ω satisfies the assumptions of Theorem 16.3,

$$\delta(U_p) = 2^{1/p'} \text{ if } 2 \leq p \leq \infty.$$

But in the case of ℓ^p,

$$\delta(U_p) = 2^{1/p} \text{ if } 2 < p < \infty$$

and

$$\delta(U_\infty) = 2.$$

Thus in general, δ provides information concerning the shape of the unit ball in a Banach space while γ does not. But, as the preceding result shows, they agree on the unit balls of L^1 and L^∞.

Theorem 16.10. *Let U denote the unit ball in $C[0,1]$ or the space M of finite complex Borel measures on $[0,1]$. Then*

$$\lambda(U) = 2.$$

Proof. Clearly $\lambda(U) \leq 2$. In the case of $C[0,1]$ it is possible, by mimicking the construction in Theorem 16.2, to construct an infinite sequence $\{x_j\}$ of continuous functions such that $\|x_j\|_\infty = 1$ (sup norm) and $\|x_j - x_k\|_\infty = 2$ $(j \neq k)$. In M consider point masses. $\square$

Theorem 16.11. *The inequality*

$$\delta(E) \leq \gamma(E) \leq 2\,\delta(E)$$

holds for any bounded subset E of a metric space.

Proof. Suppose $d' > d > 0$ and $A_1, A_2, \dots, A_n$ is a partition of E into sets of diameter $\leq d$. Then there cannot exist infinitely many points $\{x_j\}$ in E such that

$\rho(x_i, x_j) \geq d'$ $(i \neq j)$, for at least two elements of any such set would have to belong to the same A_j. Hence $\delta(E) \leq \gamma(E)$.

If $\delta(E) = 0$ then E is totally bounded and $\gamma(E) = 0$. If $0 < \delta(E) = d < d'$, there exists a finite number of balls $B(x_j, d')$, $j = 1, 2, \ldots, n$ with centers in E such that the balls $B(x_j, d')$, $j = 1, 2, \ldots, n$ cover E. Hence there is a finite partition of E into sets of diameter $\leq 2d'$. Letting $d' \downarrow d$, we get $\gamma(E) \leq 2\,\delta(E)$. $\square$

The results of Theorem 16.9 lend strong support to the conjecture that if $\delta(U) < 2$ for a Banach space X then X is reflexive. But it is not so. Kottman has observed that if X is the Banach space of R. C. James which is not reflexive, but is isometric to its second conjugate, then $\delta(U) < 2$. However, a slight variation of the conjecture is valid. We refer the reader to [50] for a proof.

Theorem 16.12. *Let X be a Banach space with unit ball U. If there exists $0 < \varepsilon < 2$ and a positive integer n such that no ε-separated subset of U has cardinality greater than n, then X is reflexive.*

Chapter V. The Extension Problem for Lipschitz-Hölder Maps between L^p Spaces

We have already shown (Theorem 11.3) that the pair (H,H), H a Hilbert space, has the extension property for Lipschitz-Hölder maps of order α, $0 < \alpha < 1$. In this chapter we treat the natural and interesting generalization of this result to L^p spaces. Starting with two σ-finite measure spaces (Ω,μ) and (N,ν) and initial values for q and p in $[1,\infty]$, the problem is to determine those values of α for which the pair $(L^q(\mu), L^p(\nu))$ has the extension property for Lip(α) maps. If $L^q_\alpha(\mu)$ denotes $L^q(\mu)$ with the norm $\|x - y\|^\alpha_q$, this is equivalent to determining those values of α for which the pair $(L^q_\alpha(\mu), L^p(\nu))$ has the extension property for contractions.

There are two obstacles to overcome in attacking this problem: first, norm inequalities must be derived to play the role in L^p analogous to that of (4.15) in the proof of Theorem 11.3; secondly, a new technique is needed to enable one to establish property (K) for the pairs $(L^q_\alpha(\mu), L^p(\nu))$. The inequalities have already been established, they are (15.1)–(15.4). And the proper relation between these inequalities and property (K) is contained in a general point-by-point extension process of Minty [63] which we now present.

§17. K-Functions and an Extension Procedure for Bilinear Forms

It will be convenient to let P_n denote the set of probability vectors in $\mathbb{R}^n$, that is, all $\xi = (\xi_1,\xi_2,\ldots,\xi_n)$ such that $\xi_j \geq 0$ and $\Sigma_j\, \xi_j = 1$.

Definition 17.1. *Let X be a set and Y a linear space over the reals. A function ϕ from $X \times X \times Y$ to $\mathbb{R}$ is called a K-function provided that*

$$\phi(x_1,x_2\,;y) \text{ is convex in } y \text{ for each pair } x_1,x_2 \in X \tag{17.1}$$

and,

for every finite sequence $(x_j,y_j) \in X \times Y$ $(j=1,2,\ldots,n)$, every $\xi \in P_n$, and every $x \in X$,

$$\sum_{j,k=1}^n \xi_j\xi_k\, \phi(x_j,x_k\,;y_j - y_k) \geq 2 \sum_{j=1}^n \xi_j\, \phi(x_j,x\,;y_j - \sum_{k=1}^n \xi_k y_k). \tag{17.2}$$

The relevance of this definition to the chapter title is not apparent. Hence it may be useful to note that if $X = H$ and $Y = H$ and

$$\phi(x_1,x_2\,;y) = \|y\|^2 - \|x_1 - x_2\|^{2\alpha} \quad (0 < \alpha \leq 1),$$

then ϕ is a K-function on $(H \times H) \times H$. The first condition is satisfied since $\|y\|^2$ is a convex function. For the second, take $\gamma = 1$, $p = 2$ and replace x_j by $x_j - x$ in (15.1) to get

$$\sum_{j,k=1}^{n} \xi_j \xi_k \|x_j - x_k\|^{2\alpha} \le 2 \sum_{j=1}^{n} \xi_j \|x_j - x\|^{2\alpha},$$

and $\gamma = 1$, $p = 2$ and $\beta = 2$ in (15.3) to get

$$\sum_{j,k=1}^{n} \xi_j \xi_k \|y_j - y_k\|^2 \ge 2 \sum_{j=1}^{n} \xi_j \left\|y_j - \sum_{k=1}^{n} \xi_k y_k\right\|^2.$$

Subtraction gives

$$\sum_{j,k=1}^{n} \xi_j \xi_k \left[\|y_j - y_k\|^2 - \|x_j - x_k\|^{2\alpha}\right] \ge 2 \sum_{j=1}^{n} \xi_j \left[\left\|y_j - \sum_{k=1}^{n} \xi_k y_k\right\|^2\right.$$
$$\left. - \|x_j - x\|^{2\alpha}\right],$$

from which it is clear that $\phi(x_1, x_2; y)$ also satisfies condition (17.2).

The next lemma will be needed in what follows.

Lemma 17.2. (Minimax Theorem of von Neumann). *Let $A, B \subset \mathbb{R}^n$ be two compact and convex sets and suppose $\Phi(\xi, \eta)$ is a function from $A \times B$ into $\mathbb{R}$ such that $\Phi(\xi, \eta)$ is convex and lower semicontinuous in η for each fixed ξ, and concave and upper semicontinuous in ξ for each fixed η. Then there exists $\xi^\circ \in A$ and $\eta^\circ \in B$ such that*

$$\Phi(\xi, \eta^\circ) \le \Phi(\xi^\circ, \eta^\circ) \le \Phi(\xi^\circ, \eta) \qquad (\xi \in A, \eta \in B).$$

Elementary proofs of this theorem will be found in the books of Berge and Ghoula-Houri [5], and Karlin [42].

Theorem 17.3. *Suppose X and Y are as in Definition 17.1 and ϕ is a K-function. If the pairs $(x_j, y_j) \in X \times Y$ $(j = 1, 2, \ldots, n)$ are such that*

$$\phi(x_j, x_k; y_j - y_k) \le 0 \qquad (j, k = 1, 2, \ldots, n) \tag{17.3}$$

and x is any element in X, then there exists an element y in the convex hull of $\{y_1, y_2, \ldots, y_n\}$ such that

$$\phi(x_j, x; y_j - y) \le 0 \qquad (j = 1, 2, \ldots, n). \tag{17.4}$$

Proof. We define a function $\Phi(\xi, \eta)$ from $P_n \times P_n$ to $\mathbb{R}$ by

$$\Phi(\xi, \eta) = \sum_{j=1}^{n} \xi_j \phi\left(x_j, x; y_j - \sum_{k=1}^{n} \eta_k y_k\right).$$

Clearly P_n is convex and compact. Also $\Phi(\xi, \eta)$ is a convex, hence a lower semi-continuous function of η, and a linear and continuous function of ξ. According to the Minimax Theorem there exists points $\xi^\circ, \eta^\circ \in P_n$ such that

$$\Phi(\xi, \eta^\circ) \le \Phi(\xi^\circ, \eta^\circ) \le \Phi(\xi^\circ, \eta) \qquad (\xi, \eta \in P_n). \tag{17.5}$$

It follows from (17.3) and (17.2) that

$$\Phi(\xi^\circ, \xi^\circ) = \sum_{j=1}^{n} \xi_j^\circ \phi(x_j, x; y_j - \Sigma \xi_k^\circ y_k) \le \frac{1}{2} \sum_{j,k=1}^{n} \xi_j^\circ \xi_k^\circ \phi(x_j, x_k; y_j - y_k) \le 0,$$

whence $\Phi(\xi, \eta^\circ) \le 0$ for all $\xi \in P_n$. The inequalities (17.4) now follow, with $y =$

$\Sigma_{j=1}^{n} \eta_{j}^{\circ} y_{j}$ by successively taking $\xi = (\delta_{ij})$, that is, one of elements of the standard basis $e_{1}, e_{2}, ..., e_{n}$ for $\mathbb{R}^{n}$. $\square$

It is the absence of linearity assumptions which makes K-functions especially suitable for extension problems involving metric spaces. In quite another direction and in connection with his work on monotone operators, Minty has recently developed an extension process for bilinear forms. His basic idea, set forth in [64], may be understood through the following result.

Theorem 17.4. *Let X and Y be real linear spaces, $\langle x,y \rangle$ a bilinear form on $X \times Y$ and $K: Y \to X$ a linear map whose associated quadratic form $\langle Ky,y \rangle$ is positive semidefinite: $\langle Ky,y \rangle \geq 0$ for all $y \in Y$. Then for every sequence $\{(x_{i},y_{i}): i=1,2,...,n\}$ in $X \times Y$ such that*

$$\langle x_{i} - x_{j}, y_{i} - y_{j} \rangle \geq 0 \quad for \quad 1 \leq i, j \leq n, \tag{17.6}$$

there exist nonnegative η_{i}° with $\Sigma_{i=1}^{n} \eta_{i}^{\circ} = 1$ such that

$$\langle x_{i} + Ky^{\circ}, y_{i} - y^{\circ} \rangle \geq 0 \quad (i = 1,2,...,n), \tag{17.7}$$

where $y^{\circ} = \displaystyle\sum_{i=1}^{n} \eta_{i}^{\circ} y_{i}$ and, if only $\displaystyle\sum_{i=1}^{n} \eta_{i}^{\circ} \leq 1$, then also

$$\sum_{i=1}^{n} \eta_{i}^{\circ} \langle x_{i}, y_{i} \rangle + \langle Ky^{\circ}, y^{\circ} \rangle \leq 0. \tag{17.8}$$

Proof. Let $D_{n} = \{\xi = (\xi_{1}, \xi_{2}, ..., \xi_{n}) \in \mathbb{R}^{n}: \xi_{i} \geq 0 \text{ and } \Sigma_{i} \xi_{i} \leq 1\}$ and define $\Phi: D_{n} \times D_{n} \to \mathbb{R}$ by

$$\Phi(\xi,\eta) = \sum_{i=1}^{n} \xi_{i}[\langle x_{i}, y_{i} \rangle - \langle x_{i}, y \rangle + \langle Ky, y_{i} \rangle - \langle Ky, y \rangle] \tag{17.9}$$

$$+ (\xi^{*} - 1) \Big[\sum_{i=1}^{n} \eta_{i} \langle x_{i}, y_{i} \rangle + \langle Ky, y \rangle \Big],$$

where $y = \Sigma_{j=1}^{n} \eta_{j} y_{j}$ and $\xi^{*} = \Sigma_{i=1}^{n} \xi_{i}$. It is readily seen that $\Phi(\xi,\eta)$ is continuous, convex in ξ, and concave in η inasmuch as the quadratic term in η is the negative semidefinite quadratic form $-\langle Ky,y \rangle$. Hence we may apply the Minimax Theorem to obtain points $\xi^{\circ}, \eta^{\circ} \in D_{n}$ such that

$$\Phi(\xi^{\circ},\eta) \leq \Phi(\xi^{\circ},\eta^{\circ}) \leq \Phi(\xi,\eta^{\circ}) \quad (\xi,\eta \in D_{n}). \tag{17.10}$$

Putting $\xi = \xi^{\circ}$ and $\eta = \xi^{\circ}$ in (17.9), we find that

$$\Phi(\xi^{\circ},\xi^{\circ}) = \frac{1}{2} \sum_{i,j=1}^{n} \xi_{i} \xi_{j} \langle x_{i} - x_{j}, y_{i} - y_{j} \rangle \geq 0,$$

the last inequality coming from (17.6). In view of (17.10) this means that $0 \leq \Phi(\xi,\eta^{\circ})$ for all $\xi \in D_{n}$. Put $y^{\circ} = \Sigma \eta_{j}^{\circ} y_{j}$. The inequalities (17.7) result by successively taking $\xi = (\delta_{ij})$ in (17.9), and (17.8) follows from (17.9) when $\xi = 0$. The same argument with $\xi^{*} = 1$ gives (17.6) with $\Sigma \eta_{i}^{\circ} = 1$. $\square$

These theorems have important corollaries. First we mention the result of Kirszbraun [44] and Valentine [86] which has already been established in the course of proving Theorem 11.3.

Corollary 17.5.   *Let* $\{(x_i,y_i):i=1,2,\ldots,n\}$ *be a sequence in* $H \times H$ *such that* $\|y_i - y_j\| \leq \|x_i - x_j\|$ $(1 \leq i,\ j \leq n)$. *Then for every* $x \in H$ *there exists a* y *in the convex hull of* $\{y_1,y_2,\ldots,y_n\}$ *such that*

$$\|y_i - y\| \leq \|x_i - x\| \qquad (i = 1,2,\ldots,n).$$

Proof.   Apply Theorem 17.2 with $X = Y = H$, $\langle , \rangle$ the inner product in H, and $\phi(x_1,x_2;y) = \|y\|^2 - \|x_1 - x_2\|^2$. $\square$

The following theorem of B. Grünbaum [31] follows directly from Theorem 17.3.

Corollary 17.6.   *If the sequence* $\{(x_i,y_i): i=1,2,\ldots,n\}$ *in* $H \times H$ *satisfies the condition* $(x_i - x_j,\ y_i - y_j) \geq 0$ *for* $1 \leq i,\ j \leq n$ *and* $\lambda > 0$, *there exists a* y° *in the convex hull of* $\{y_1,y_2,\ldots,y_n\}$ *such that*

$$(x_i + \lambda y^\circ,\ y_i - y^\circ) \geq 0 \qquad (i=1,2,\ldots,n). \tag{17.11}$$

The natural culmination of these results is a fundamental theorem of Minty on maximal monotone sets [60], [70].

Definition 17.7.   *A subset* A *of* $H \times H$ *is monotone provided that*

$$(x_1 - x_2, y_1 - y_2) \geq 0 \quad \text{for all pairs} \ (x_i,y_i) \in A \quad (i = 1,2),$$

and a monotone set which is not properly contained in any other monotone set is called maximal monotone.

Theorem 17.8.   *Let* A *be a monotone subset of* $H \times H$. *If for some* $\lambda > 0$,

$$\{x + \lambda y : (x,y) \in A\} = H, \tag{17.12}$$

then A *is maximal monotone. If* A *is maximal monotone then* (17.12) *holds for every* $\lambda > 0$.

Proof.   Let (17.12) hold for some $\lambda > 0$. If (u,v) is such that

$$(x - u, y - v) \geq 0 \quad \text{for all} \quad (x,y) \in A, \tag{17.13}$$

then choose $(w,z) \in A$ such that

$$w + \lambda z = u + \lambda v.$$

Replacing (x,y) by $(w,z\ )$ in (17.13) gives $\lambda(v - z, z - v) \geq 0$. Hence $v = z$ and $w = u$, that is, $(u,v) \in A$ and A is maximal.

Let A be maximal monotone, $\lambda > 0$ and $w \in H$. It suffices to prove the existence of an element $u \in H$ such that

$$(x-u,\ y + \frac{1}{\lambda}(u-w)) \geq 0 \quad \text{for all} \quad (x,y) \in A, \tag{17.14}$$

for by maximality the pair $(u, -\frac{1}{\lambda}(u-w))$ must belong to A. Putting $v = -\frac{1}{\lambda}(u-w)$, it follows that $w = u + \lambda v$ where $(u,v) \in A$.

It is enough to establish (17.14) with $w = 0$, since A is maximal monotone if and only if the set $\{(x, y - \frac{1}{\lambda}w):(x,y) \in A\}$ is maximal monotone. According to

Corollary 17.6 the set

$$U(x,y) = \{u \in H : (x-u, \, y + \frac{1}{\lambda} u) \geq 0\}$$

is nonempty for every pair $(x,y) \in A$ and the sets $\{U(x,y) : (x,y) \in A\}$ possess the finite intersection property. Moreover each $U(x,y)$ is bounded and weakly closed, hence weakly compact. The existence of an element $u \in H$ satisfying (17.14) therefore follows and the proof is complete. $\square$

For a discussion of the coercivity condition (17.8) see [64]. For an introduction to the general theory of monotone operators and their applications to existence theorems for integral equations and certain elliptic partial differential equations, see the survey article of Minty [62] and the works of Browder [13] and Pazy [70].

§18. Examples of K-Functions

Let ϕ be a K-function on $X \times X \times Y$. In case Y is trivial we write

$$\phi(x_1, x_2; y) = -k_1(x_1, x_2) \qquad (x_1, x_2 \in X).$$

Then (17.1) is vacuous and inequality (17.2) reduces to

$$\sum_{j,k=1}^{n} \xi_j \xi_k \, k_1(x_j, x_k) \leq 2 \sum_{j=1}^{n} \xi_j \, k_1(x_j, x) \tag{18.1}$$

for all $x, x_1, x_2, \ldots, x_n$ in X and $\xi \in P_n$ $(n \geq 1)$.

If X is empty, put

$$\phi(x_1, x_2; y) = k_2(y) \qquad (y \in Y).$$

In this case conditions (17.1) and (17.2) become

$$k_2(y) \text{ is a convex function of } y \text{ and}$$

$$\sum_{j,k=1}^{n} \xi_j \xi_k \, k_2(y_j - y_k) \geq 2 \sum_{j=1}^{n} \xi_j \, k_2(y_j - \sum_{k=1}^{n} \xi_k y_k) \tag{18.2}$$

for all $y_1, y_2, \ldots, y_n$ in Y and all $\xi \in P_n$ $(n \geq 1)$.

In the opposite direction—and this is a recipe for constructing K-functions— the function

$$\phi(x_1, x_2; y) = k_2(y) - k_1(x_1, x_2)$$

will be a K-function on $X \times X \times Y$ whenever k_1 is a function on $X \times X$ satisfying (18.1) and k_2 is a function on Y satisfying (18.2). But not every K-function arises in this manner for clearly, in case X is also a linear space, a bilinear form on $X \times Y$ is a K-function.

It is easy to see that (18.1) is satisfied in any one of the following situations:
(1) $k_1(x_1, x_2) =$ a nonnegative constant.
(2) $k_1(x_1, x_2)$ is a semi-metric on X.
(3) $k_1(x_1, x_2) = \rho(x_1, x_2)^{\alpha}$ $(0 < \alpha \leq 1)$ where ρ is a metric for X.
(4) $k_1(x_1, x_2) = F \circ \rho(x_1, x_2)$ where ρ is a metric for X and $F \in N(X)$.

(5) $k_1(x_1,x_2) = \|x_1 - x_2\|^\alpha$ where $(0 < \alpha \leq 2)$ and X is a Hilbert space (see 4.15).

(6) $k_1(x_1,x_2) = g \circ \rho$ where ρ is a metric on X and g is a real-valued function on $\mathbb{R}^+$ such that $g(0) = 0$, $g(t) > 0$ for $t > 0$, $g(t)$ is subadditive.

(7) $k_1(x_1,x_2) = \|x_1 - x_2\|_p^p$ for $1 < p \leq 2$ and $X = L^p(\mu)$.

And condition (18.2) is satisfied by each of the following:

(8) $k_2(y) = \|y\|_p^p$ for $2 \leq p < \infty$ and $Y = L^p(v)$.

(9) $k_2(y)$ is a linear functional on Y.

(10) $k_2(y) = y^4$ and $y = \mathbb{R}$.

The assertion in (10) follows from (8) or from the identity

$$\Sigma \, \xi_i \xi_j \, |y_i - y_j|^4 = 2 \, \Sigma_i \, \xi_i \, |y_i - y|^4 + 6 \, (\Sigma_i \, \xi_i y_i^2 - y^2)^2,$$

where $y = \Sigma_i \, \xi_i y_i$ and $\xi \in P_n$.

Several writers, unaware of Schoenbergs' early work, have given new proofs for (5). Brezis [12], and Wells and Hayden [37] used the M. Riesz convexity theorem, but the same idea was used much earlier in [38]. An elementary proof of a somewhat stronger inequality is due to Fox:

$$\Sigma \, \xi_i \xi_j \, \|x_i - x_j\|^{2\sigma} \leq \Sigma \, \xi_i \xi_j \, (\|x_i\|^2 + \|x_j\|^2)^\alpha \quad (0 < \alpha \leq 1).$$

It is a direct consequence of the following self-evident lemma of Moser and Minty [63].

Lemma 18.1. *If* $x_1, x_2, \ldots, x_n$ *are points in a real inner product space,* $\xi_i > 0$ *($1 \leq i \leq n$) and* $\beta > 0$ *then*

$$\sum_{i,j} \frac{(x_i, x_j)}{(\xi_i + \xi_j)^\beta} = \Gamma(\beta)^{-1} \int_0^\infty \| \, \Sigma_i \, e^{-\xi_i t} \, x_i \|^2 \, t^{\beta - 1} \, dt.$$

§19. The Contraction Extension Problem for the Pairs (L_α^q, L^p)

Theorem 19.1. *The pair* $(L_\alpha^q(\mu), L^p(v))$ *has the extension property for contractions in each of the following cases:*

(i) $2 \leq p < \infty$, $1 < q \leq 2$ and $0 < \alpha \leq q/p$.

(ii) $2 \leq p < \infty$, $2 \leq q < \infty$ and $0 < \alpha \leq q'/p$.

(iii) $1 < p \leq 2$, $1 < q \leq 2$ and $0 < \alpha \leq q/p'$.

(iv) $1 < p \leq 2$, $2 \leq q < \infty$ and $0 < \alpha \leq q'/p'$.

If X *is a metric space, then the pair* $(X, L^p(v))$ *has the extension property for* Lip(α) *maps if* $0 < \alpha \leq 1/p$ *and* $2 \leq p < \infty$, *or if* $0 < \alpha \leq 1/p'$ *and* $1 < p \leq 2$.

Proof. We consider case (i) and let f be a Lipschitz-Hölder map of order α from a subset D of $L^q(\mu)$ into $L^p(v)$. Let $\beta = q/\alpha$ and note that $\beta \geq p \geq 2$. The K-function associated with this case is the map

$$\phi : L^q(v) \times L^q(v) \times L^p(\mu) \to \mathbb{R}$$

defined by

$$\phi(x_1, x_2; y) = \|y\|_p^\beta - \|x_1 - x_2\|_q^q.$$

Clearly, $\phi(x_1, x_2; y)$ is convex in y for fixed x_1 and x_2. From the inequalities (15.1) and (15.4) with $\gamma = 1$, it follows that $\phi(x_1, x_2; y)$ satisfies the second condition (17.2) required of K-functions.

In order to show that the domain of f can be extended beyond D to any point z in its complement, it is enough to show that

$$\bigcap_{x \in D} B(f(x), \|z - x\|_q^\alpha) \neq \emptyset .$$

Fix $x_0 \in D$ and define

$$C(x) = B(f(x), \|z - x\|_q^\alpha) \cap B(f(x_0), \|z - x_0\|_q^\alpha)$$

for each $x \in D$. Clearly, $\bigcap_{x \in D} C(x) = \bigcap_{x \in D} B(f(x), \|z - x\|_q^\alpha)$. Since balls are weakly compact in the reflexive space $L^p(v)$, we may conclude that $\bigcap_{x \in D} C(x) \neq \emptyset$ provided the collection $\{C(x) : x \in D\}$ enjoys the finite intersection property. To establish this, choose a finite set $x_1, x_2, \ldots, x_n$ in D and note that it follows from the inequalities

$$\|f(x_j) - f(x_k)\|_p^\beta \leq \|x_j - x_k\|_q^{\alpha\beta} = \|x_j - x_k\|_q^q \quad (0 \leq j, k \leq n)$$

that

$$\phi(x_i, x_j; f(x_i) - f(x_j)) \leq 0 \qquad (0 \leq i, j \leq n).$$

By Theorem 17.3 there exists a $y \in L^p(v)$ such that $\phi(x_i, z; f(x_i) - y) \leq 0$ for $i = 0, 1, \ldots, n$. Therefore,

$$\|f(x_i) - y\|_p^\beta - \|x_i - z\|_q^q \leq 0 \quad \text{or} \quad \|f(x_i) - y\|_p \leq \|x_i - z\|_q^\alpha$$

for $i = 0, 1, \ldots, n$. Hence $y \in \bigcap_{j=}^{n} C(x_j)$ and y belongs to the convex hull of $\{f(x_0), f(x_1), \ldots, f(x_n)\}$. Thus $\bigcap_{x \in D} C(x) \neq \emptyset$, moreover a w in this intersection can be chosen in the closed convex hull of $f(D) = \{f(x) : x \in D\}$ and, by defining $f(z) = w$, we obtain $\|f(z) - f(x)\|_p \leq \|z - x\|_q^\alpha$ for all $x \in D$. A simple argument based on Zorn's lemma now shows that f can be extended to all of $L^q(\mu)$ in such a way that the Lip(α) condition is preserved and the range of the extension lies in the closed convex hull of $f(D)$.

We omit the entirely similar arguments which show that (ii) follows from (15.2) and (15.4), (iii) follows from (15.1) and (15.3), and (iv) follows from (15.2) and (15.3).

The last assertion concerning extension of Lip(α) maps from a metric space into $L^p(v)$ is proved by the same procedure. One needs only note that, by (15.3) and (15.4), the map $\phi : X \times X \times L^p(v) \to \mathbb{R}$, defined by

$$\phi(x_1, x_2; y) = \|y\|_p^\beta - \rho(x_1, x_2)$$

is a K-function, where $\beta \geq p$ for $2 \leq p < \infty$ and $\beta \geq p'$ for $1 < p \leq 2$. $\square$

Analogues of these results in Orlicz spaces are established in [16].

In general, the restriction placed on α in the various cases of the preceding theorem are necessary.

Theorem 19.2. *In each of the cases (i)–(iv) of the preceding theorem, the range of α is sharp provided the underlying spaces are infinite-dimensional.*

The proof of this theorem will follow a series of lemmas. We begin with the construction of sets in ℓ_n^p which exhibit useful geometric properties.

Corresponding to the 4×8 matrix

$$A = \begin{pmatrix} 1 & 1 & 1 & 1 & -1 & -1 & -1 & -1 \\ 1 & 1 & -1 & -1 & 1 & 1 & -1 & -1 \\ 1 & -1 & -1 & 1 & 1 & -1 & 1 & -1 \\ 1 & -1 & 1 & -1 & -1 & 1 & 1 & -1 \end{pmatrix},$$

define an operator T on the set of all matrices over $\mathbb{R}$ as follows:

If $B = (b_{ij})$ is an $n \times m$ matrix, then $T(B)$ is the $4n \times 8m$ matrix $(b_{ij}A)$ with the obvious identification between an $n \times m$ matrix of 4×8 matrices and a $4n \times 8m$ matrix. Let $X^1 = A$ and define $X^k = T(X^{k-1})$ for $k > 1$. Note that X^k is a $2^{2k} \times 2^{3k}$ matrix each of whose entries is ± 1.

We omit the straightforward arguments for the next two lemmas.

Lemma 19.3. *If $k \geq 1$, then*

$$\sum_{i=1}^{2^{2k}} X_{i,j}^k = 0 \quad \text{or} \quad \pm 2^{2k} \quad \text{for} \quad 1 \leq j \leq 2^{3k},$$

and the second possibility holds for exactly 2^k choices of j.

Lemma 19.4. *If $k \geq 1$ and $1 \leq r < s \leq 2^{2k}$, then*

$$\sum_{j=1}^{2^{3k}} |X_{r,j}^k - X_{s,j}^k|^p = 2^{3k-1} \cdot 2^p.$$

For $k \geq 1$ define the 2^{2k} vectors $z_1^k, z_2^k, \ldots, z_{2^{2k}}^k$ in $\ell_{2^{3k}}^p$ by

$$(z_i^k)_j = 2^{-3k/p} (X_{i,j}^k - 2^{-2k} \sum_{r=1}^{2^{2k}} X_{r,j}^k), \quad 1 \leq j \leq 2^{3k}.$$

Of course $(z_i^k)_j$ denotes the j-th component of z_i^k. We record the essential properties of these vectors.

Lemma 19.5. *The points $\{z_i^k\}_{i=1}^{2^k}$ satisfy the following conditions:*

(a) *Each component of z_i^k is 0 or $\pm 2^{-3k/p}$;*

(b) $\sum_{i=1}^{2^{2k}} z_i^k = 0$;

(c) $\|z_i^k\|_p = (1 - 2^{-2k})^{1/p}$;

(d) $\|z_i^k - z_j^k\|_p = 2^{1/p'}$ *for $i \neq j$; and*

(e) $\displaystyle\min_{z \in \ell_{2^{3k}}^p} \max_{1 \leq i \leq 2^{2k}} \|z - z_i^k\|_p = (1 - 2^{-2k})^{1/p}.$

Proof. Part (a) follows from Lemma 19.3 and (b) follows from the definition of z_i^k. Also, by Lemma 19.3, we see that z_i^k has 2^k zero entries, so that $\|z_i^k\|_p = 2^{-3k/p}(2^{3k} - 2^k)^{1/p} = (1 - 2^{-2k})^{1/p}$. By Lemma 19.4, $\|z_i^k - z_j^k\|_p = 2^{-3k/p}(2^{(3k-1)/p})$ (2)
$= 2^{1-1/p} = 2^{1/p'}$.

In order to prove (e), suppose there exists $z \in \ell_{2^{3k}}^p$ such that

$$\max_{1 \leq i \leq 2^{2k}} \|z - z_i^k\|_p < (1 - 2^{-2k})^{1/p} = \max_{1 \leq i \leq 2^{2k}} \|z_i^k\|_p.$$

Let F_i be the linear functional on $\ell^p_{2^{2k}}$ such that $\|F_i\| = 1$ and $F_i(z^k_i) = \|z^k_i\|_p$, for $1 \le i \le 2^{2k}$. By virtue of (a) we have

$$F_i(x) = (x, z^k_i) \, 2^{3k/p} \cdot 2^{-3k(p-1)/p} (1 - 2^{-2k})^{-1/p'} = (x, z^k_i) \cdot \alpha$$

for $x \in \ell^p_{2^{2k}}$. Also, $F_i(z) > 0$ since

$$(1 - 2^{-2k})^{1/p} > \|z - z^k_i\|_p \ge |F_i(z - z^k_i)| \ge (1 - 2^{-2k})^{1/p} - F_i(z).$$

Hence we have the contradiction

$$0 < \sum_{i=1}^{2^{2k}} F_i(z) = \alpha \cdot \sum_{i=1}^{2^{2k}} (z, z^k_i) = \alpha \cdot \left(z, \sum_{i=1}^{2^{2k}} z^k_i\right) = 0.$$

This proves (e). $\square$

Let $e^k_1, e^k_2, \ldots, e^k_{2^{2k}}$ denote the 2^{2k} unit vectors in the standard basis for $\ell^p_{2^{2k}}$, that is, $(e^k_i)_j = \delta_{ij}$ for $1 \le j \le 2^{2k}$.

Lemma 19.6. *The following hold for* $\{e^k_i\}$:

(f) $\|e^k_i - e^k_j\|_p = 2^{1/p}$, $i \ne j$;
(g) $\|e^k_i\|_p = 1$;
(h) $\lim\limits_{k \to \infty} \min\limits_{y \in \ell^p_{2^{2k}}} \max\limits_{1 \le i \le 2^{2k}} \|e^k_i - y\|_p = 1$.

Proof. Only (h) requires argument. For each $k \ge 1$ the min max is assumed at some point $y = (y_1, y_2, \ldots, y_{2^{2k}})$ for which $0 \le y_i$. Choose s so that $y_s \le y_j$ for $1 \le j \le 2^{2k}$. Then

$$\max_{1 \le j \le 2^{2k}} \|y - e^k_j\|_p \ge \Big\{ (1 - y_s)^p + \sum_{j \ne s} y^p_j \Big\}^{1/p}$$

$$\ge \{ (1 - y_s)^p + (2^{2k} - 1) y^p_s \}^{1/p}.$$

Now the function $(1 - t)^p + (2^{2k} - 1) t^p$ assumes it minimum for positive t at $t = [(2^{2k} - 1)^{1/(p-1)} + 1]^{-1} = c^{-1}_k$, so if we define $w^k = (c^{-1}_k, c^{-1}_k, \ldots, c^{-1}_k)$, then

$$\min_{y \in \ell^p_{2^{2k}}} \max_{1 \le i \le 2^{2k}} \|e^k_i - y\|_p = \|w^k - e^k_i\|_p = \{ (1 - c^{-1}_k)^p + (2^{2k} - 1) c^{-p}_k \}$$

and this expression tends to 1 as $k \to \infty$. $\square$

Proof of Theorem 19.2. Suppose the pair $(L^q_\alpha(\mu), L^p(\nu))$ has the extension property for contractions, where $1 < p$, $q \le 2$. Let $d = 2^{1/\alpha p' - 1/q}$ and set $D = \{ de^k_i : 1 \le i \le 2^{2k} \}$. Define $f : D \to \ell^p_{2^{2k}}$ by $f(de^k_i) = z^k_i$. Then

$$\|f(de^k_i) - f(de^k_j)\|_p = \|z^k_i - z^k_j\|_p = 2^{1/p'} = d^\alpha 2^{\alpha/q}$$
$$= d^\alpha \|e^k_i - e^k_j\|^\alpha_q = \|de^k_i - de^k_j\|^\alpha_q,$$

so $f \in \mathrm{Lip}(\alpha)$. By assumption the map f can be extended to 0 so as to preserve the Lipschitz condition. According to Lemma 19.5 (e), the best choice for $f(0)$ is 0. Therefore it must be true that $\|z^k_i\|_p \le \|de^k_i\|^\alpha_q$, that is, $(1 - 2^{-2k})^{1/p} \le 2^{1/p' - \alpha/q}$ and this must hold for all integers k since we are assuming that all pairs $(L^q_\alpha(\nu), L^p(\mu))$ have the extension property for contractions. Hence $2^{1/p' - \alpha/q} \ge 1$ or $q/p' \ge \alpha$.

For $1 < p \le 2 \le q < \infty$ define f by $f(2^{1/\alpha p' - 1/q'} z^k_i) = z^k_i$; for $2 \le p$, $q < \infty$ define f by $f(2^{1/\alpha p - 1/q'} z^k_i) = e^k_i$; and for $1 < q \le 2 \le p < \infty$ define f by $f(2^{1/\alpha p - 1/q} e^k_i) = e^k_i$. In each case extension of f to 0 implies sharpness in the corresponding part of Theorem 19.1. $\square$

Bibliography

1. Akhiezer, N.: The classical moment problem. Edingurgh: Oliver and Boyd 1965.
2. Arens, R., Eells, J., Jr.: On embedding uniform and topological spaces. Pacific J. Math. **6**, 397–403 (1956).
3. Aronszajn, N., Panitchpakdi, P.: Extension of uniformly continuous transformations and hyperconvex metric spaces. Pacific J. Math. **6**, 405–439 (1956).
4. Banach, S.: Théorie des opérations linéaires. Monografje Matematyczne, vol. 1, Warsaw 1932.
5. Berge, C., Ghoula-Houri, A.: Programming, games and transportation networks. New York: Wiley 1965.
6. Blichfeldt, H.: The minimum value of quadratic forms, and the closest packing of spheres. Math. Ann. **101**, 605–608 (1929).
7. Blumenthal, L.: New theorems and methods in determinant theory. Duke Math. J. **2**, 396–404 (1936).
8. Blumenthal, L.: A note on the four point property. Bull. Amer. Math. Soc. **39**, 423–426 (1933).
9. Blumenthal, L.: Remarks concerning the euclidean four-point property. Ergebnisse eines math. Koll. (Vienna) **7**, 8–10 (1936).
10. Bochner, S.: Stable laws of probability and completely monotone functions. Duke Math. J. **3**, 726–728 (1937).
11. Bretagnolle, J., Dachuna Castelle, D., Krivine, J.: Lois stables et espaces L^p. Ann. Inst. H. Poincare Sect. B (3)II, 231–259 (1966).
12. Brezis, H.: Prolongement d'applications Lipschitziennes et de semi-groupes de contractions. Séminaire Choquet: Initiation à l'analyse, 9 e anné, n° 19 (1969/70).
13. Browder, F.: Nonlinear functional analysis and nonlinear integral equations of the Hammerstein and Urysohn type. Contributions to nonlinear functional analysis: proceedings. E. Zarantonello ed. New York: Academic Press 1971.
14. Burlak, J., Rankin, R., Robertson, A.: The packing of spheres in the space ℓ_p. Proc. Glasgow Math. Assoc. **4**, 22–25 (1958).
15. Chakerian, G., Klamkin, M.: Inequalities for sums of distances. Amer. Math. Monthly **80**, 1009–1017 (1973).
16. Cleaver, C.: On the extension of Lipschitz-Hölder maps on Orlicz spaces. Studia Math. **42**, 195–204 (1972).
17. Czipszer, J., Geher, L.: Extensions of functions satisfying a Lipschitz condition. Acta. Math. Acad. Sci. Hungar **6**, 213–219 (1955).
18. Danzer, L., Grünbaun, B., Klee, V.: Helly's theorem and its relatives. In: Proceedings of a symposium on pure math. **7**, pp. 101–180. Providence, R.I.: Amer. Math. Soc. 1963.
19. Darbo, G.: Punti uniti in transformazioni a codominio non compatto. Padua Univ. Sem. Math. **24**, 84–92 (1955).
20. Day, M.: Some characterizations of inner-product spaces. Trans. Amer. Math. Soc. **62**, 320–337 (1947).
21. Debrunner, H., Flor, P.: Ein Erweiterungssatz für Monotone mengen. Arch. Math. **15**, 445–447 (1964).
22. Dunford, N., Schwartz, J.: Linear operators I. New York: Interscience 1958.
23. Edelstein, M.: On nonexpansive mappings. Proc. Amer. Math. Soc. **15**, 689–695 (1964).
24. Edelstein, M.: On non-expansive mappings of Banach spaces. Proc. Cambridge Philos. Soc. **60**, 439–447 (1964).
25. Edelstein, M.: Thompson, A.: Contractions, isometries and some properties of inner-product spaces. Nederl. Akad. Wetensch. Proc. Ser. A. **70**, 326–331 (1967).

26. Einhorn, S.: Functions positive definite on C[0,1]. Proc. Amer. Math. Soc. **22**, 702–703 (1969).
27. de Figueiredo, D., Karlovitz, L.: On the radial projection in normed spaces. Bull. Amer. Math. Soc. **73**, 364–368 (1967).
28. de Figueiredo, D., Karlovitz, L.: On the projection onto convex sets and the extension of contractions. In: Proceedings of a Conference on Projections and Related Topics. Clemson Univ. 1967.
29. de Figueiredo, D., Karlovitz, L.: On the extension of contractions on normed spaces. In: Proceedings of Symposia in Pure Mathematics, v.18 pt.1. Providence, R.I.: Amer. Math. Soc. 1970.
30. Gikhman, I., Skorokhod, A.: Introduction to the theory of random processes. Philadelphia: W. B. Saunders Co. 1969.
31. Grünbaum, B.: A generalization of theorems of Kirszbraun and Minty. Proc. Amer. Math. Soc. **13**, 812–814 (1962).
32. Grünbaum, B.: On a theorem of Kirszbraun. Bull. Res. Counc. Israel **7F**, 129–132 (1958).
33. Grümbaum, F., Zarantonello, E.: On the extension of uniformly continuous mappings. Michigan Math. J. **15**, 65–74 (1968).
34. Halmos, P.: Measure theory. New York: Van Nostrand 1959.
35. Hanner, O.: Intersections of translates of convex bodies. Math. Scand. **4**, 65–87 (1956).
36. Harris, L.: Bounds on the derivatives of holomorphic functions of vectors. Proc. Colloq. Analysis, Rio de Janeiro, 1972 (L. Nachbin, ed.) Act. Sci. et Ind., Paris: Hermann 1973.
37. Hayden, T., Wells, J.: On the extension of Lipschitz-Hölder maps of order α. J. Math. Anal. Appl. **33**, 627–640 (1971).
38. Hua, L.: A remark on a result due to Blichfeldt. Bull. Amer. Math. Soc. **51**, 537–539 (1945).
39. Jameson, G.: Ordered linear spaces, Lecture Notes in Math. no.141. Berlin-Heidelberg-New York: Springer 1970.
40. Kakutani, S.: Concrete representation of abstract (L)-spaces and the mean ergodic theorem. Ann. of Math. **42**, 523–537 (1941).
41. Kakutani, S.: Some characterizations of Euclidean space. Japan J. Math. **16**, 93–97 (1939).
42. Karlin, S.: Mathematical methods and theory in games, programming, and economics, v.1. Reading, Mass.: Addison-Wesley 1962.
43. Kelley, J.: Banach spaces with the extension property. Trans. Amer. Math. Soc. **72**, 323–326 (1952).
44. Kirszbraun, M.: Über die zusammenziehende und Lipschitzsche transformationen. Fund. Math. **22**, 77–108 (1934).
45. Klee, V.: On a theorem of Bela Sz.-Nagy. Amer. Math. Monthly **60**, 618–619 (1953).
46. Klee, V.: On certain intersection properties of convex sets. Canad. J. Math. **3**, 272–275 (1951).
47. Kolmogorov, A., Gnedenko, B.: Limit distributions for sums of independent random variables. Reading, Mass.: Addison-Wesley 1954.
48. Kolmogorov, A., Tihomirov, V.: ε-entropy and ε-capacity of sets in function spaces. Uspehi Mat. Nauk. **14** no.2, 3–86 (1959). [English transl.] Amer. Math. Soc. Transl. (2) **17**, 277–364 (1961).
49. Köthe, G.: Topologische lineare Räume. Berlin-Heidelberg-New York: Springer 1966.
50. Kottman, C.: Packing and reflexivity in Banach spaces. Trans. Amer. Math. Soc. **150**, 565–576 (1970).
51. Krein, M. G.: Sur le probleme du prolongement des function hermitiennes positive and continues. C.R. (Doklady) Acad. Sci. USSR (N.S.) **26**, 17–22 (1940).
52. Kuelbs, J.: Positive definite symmetric functions on linear spaces. J. Math. Anal. Appl. **42**, 413–426 (1973).
53. Lindenstrauss, J.: Extension of compact operators. Memoirs of Amer. Math. Soc. No. 48 (1964).
54. Loeve, M.: Probability theory. New York: Van Nostrand 1963.
55. Lorch, E.: On certain implications which characterize Hilbert space. Ann. of Math. **49**, 523–532 (1948).
56. McShane, E. J.: Extension of the range of functions. Bull. Amer. Math. Soc. **40**, 837–842 (1934).
57. Menger, K.: Die Metrik des Hilbert-raumes. Akad. Wiss. Wien Abh. Math.-Natur. Kl. **65**, 159–160 (1928).
58. Menger, K.: New foundations of euclidean geometry. Amer. J. Math. **53**, 721–745 (1931).
59. Mickle, E.: On the extension of a transformation. Bull. Amer. Math. Soc. **55**, 160–164 (1949).
60. Minty, G.: Monotone (nonlinear) operators in Hilbert space. Duke Math. J. **29**, 341–346 (1962).

61. Minty, G.: On a "monotonicity" method for the solution of nonlinear equations in Banach spaces. Proc. Nat. Acad. Sci. U.S.A. **50**, 1038–1041 (1963).
62. Minty, G. On some aspects of the theory of monotone operators, theory and applications of monotone operators. In: Proceedings of a NATO Advanced Study Institute in Venice, Italy, pp. 67–82 (1968).
63. Minty, G.: On the extension of Lipschitz, Lipschitz-Hölder and monotone functions. Bull. Amer. Math. Soc. **76**, 334–339 (1970).
64. Minty, G.: A finite-dimensional tool-theorem in monotone operator theory. Advances in Math. **12**, no.1, 1–7 (1974).
65. Nachbin, L.: A theorem of the Hahn-Banach type for linear transformations. Trans. Amer. Math. Soc. **68**, 28–46 (1950).
66. von Neumann, J., Schoenberg, I.: Fourier integrals and metric geometry. Trans. Amer. Math. Soc. **50**, 226–251 (1941).
67. Neveu, J.: Mathematical foundations of the calculus of probability. San Francisco: Holden-Day 1965.
68. Nussbaum, A.: Radial exponentially convex functions. J. Analyse Math. **25**, 277–288 (1972).
69. Nussbaum, R.: The fixed point index for local condensing maps. Ann. Mat. Pura Appl. (4) **89**, 217–258 (1971).
70. Pazy, A.: Semi-groups of nonlinear contractions in Hilbert space. In: Problems in Non-Linear Analysis. Rome: C.I.M.E. 1970.
71. Phelps, R.: Lectures on Choquet's theorem. Princeton: Van Nostrand 1966.
72. Polya, G., Szego, G.: Aufgaben und Lehrsätze aus der Analysis, v.2. Berlin-Göttingen-Heidelberg: Springer 1954.
73. Rankin, R.: On the sums of powers of linear forms II. Ann. of Math. **50**, 699–704 (1949).
74. Rudin, W.: Fourier analysis on groups. New York-London: Interscience 1962.
75. Rudin, W.: The extension problem for positive-definite functions. Illinois J. Math. **7**, 532–539 (1963).
76. Rudin, W.: An extension theorem for positive-definite functions. Duke Math. J. **37**, 49–53 (1970).
77. Schoenberg, I.: Remarks to Maurice Frechet's article "Sur la definition axiomatique d'une classe d'espaces distancies vectoriellement applicable sur l'espace de Hilbert." Ann. of Math. (2) **36**, 724–732 (1935).
78. Schoenberg, I.: Regular simplices and quadratic forms. J. London Math. Soc. **12**, 48–55 (1937).
79. Schoenberg, I.: On certain metric spaces arising from Euclidean space by a change of metric and their imbedding in Hilbert space. Ann. of Math. (2) **38**, 787–793 (1937).
80. Schoenberg, I.: Metric spaces and positive definite functions. Trans. Amer. Math. Soc. **44**, 522–536 (1938).
81. Schoenberg, I.: Metric spaces and completely monotonic functions. Ann. of Math. (2) **39**, 811–841 (1938).
82. Schoenberg, I.: On a theorem of Kirszbraun and Valentine. Amer. Math. Monthly **60**, 620–622 (1953).
83. Schönbeck, S.: Extension of nonlinear contractions. Bull. Amer. Math. Soc. **72**, 99–101 (1966).
84. Schonbeck, S.: On the extension of Lipschitz maps. Ark. Mat. **7**, 201–209 (1967).
85. Schwartz, J.: Nonlinear functional analysis. New York: Gordon and Breach 1969.
86. Valentine, F.: A Lipschitz condition preserving extension for a vector function. Amer. J. Math. **67**, 83–93 (1945).
87. Valentine, F.: On the extension of a vector function so as to preserve a Lipschitz condition. Bull. Amer. Math. Soc. **49**, 100–108 (1943).
88. Valentine, F.: Contractions in non-euclidean spaces. Bull. Amer. Math. Soc. 50, 710–713 (1944).
89. Widder, D.: The Laplace transform. Princeton: Princeton Univ. Press 1941.
90. Williams, L., Wells, J., Hayden, T.: On the extension of Lipschitz-Hölder maps on L' spaces. Studia Math. **39**, 29–38 (1971).
91. Wilson, W.: On certain types of continuous transformations of metric spaces. Amer. J. Math. **57**, 62–68 (1935).
92. Yosida, K., Kakutani, S.: Operator-theoretical treatment of Markoff's process. Ann. of Math. **42**, 188–228 (1941).
93. Zygmund, A.: Trigonometric series II, 2nd ed. Cambridge: Cambridge University Press 1959.

Author Index

Subject Index

List of Symbols

Ergebnisse der Mathematik und ihrer Grenzgebiete